Lecture Notes in Mathematics

Edited by A. Dold, Heidelberg and B. Eckmann, Zürich

Series: Mathematisches Institut der Universität Erlangen-Nü

Advisers: H. Bauer und K. Jacobs

355

Maurice Sion

University of British Columbia, Vancouver/Canada

A Theory of Semigroup Valued Measures

Springer-Verlag

Berlin · Heidelberg · New York 1973

AMS Subject Classifications (1970): 28-02, 28 A 15, 28 A 45, 28 A 50, 46 G 10, 46 G 15

ISBN 3-540-06542-3 Springer-Verlag Berlin · Heidelberg · New York
ISBN 0-387-06542-3 Springer- Verlag New York · Heidelberg · Berlin

Offsetdruck: Julius Beltz, Hemsbach/Bergstr.

PREFACE

This monograph contains mostly material developed by the author during the past several years, much of it presented in seminar lectures at the University of British Columbia. It is based on a mimeographed set of notes entitled "Lectures on Vector-Valued Measures" issued in 1969-70.

It is somewhat ironic that the original motivation for developing this theory has not been included here. It involved a problem on diffusion of measures of relevance in potential theory. It led to an integral representation theorem for vector-valued measures and from there to a reexamination of the notions of measurable function and integral. Thus, the order in which the material is presented here is the reverse of the order in which it was developed. It is unfortunate that the applications, which constitute the more interesting aspects of the theory, have been omitted in order not to delay publication of these results any further. We plan to issue a sequel containing applications in the near future and hope that the few indications given here and there will enable at least those more familiar with the field to appreciate the usefulness of the theory.

M. Sion

TABLE OF CONTENTS

CHAPTER I

SEMIGROUP VALUED OUTER MEASURES

The essence of the Carathéodory approach to measure theory is to generate, from some given limited data, a function μ defined from the outset on the family of all subsets of a space S and then to study various approximation and additivity properties of μ. In this approach, the family of measurable sets is not picked out in advance, but is determined by μ in a very natural way. The spotlight, therefore, is on the function μ itself, or the particular process used for generating it, rather than on its behavior on some special family chosen a priori. The advantage of this approach is that it permits one to maintain, without loss of naturalness and simplicity, a consistent and unified point of view in considering such diverse topics as extension of a measure, integral, capacity, topological measures, cylinder measures, etc.

We shall follow an approach very similar in spirit to that of Carathéodory in our development of a theory for measures taking values in a topological semigroup. No order considerations are involved however.

0. Preliminaries.

Throughout this chapter,

S is an abstract space,

H is a family of subsets of S,

X is a commutative, Hausdorff, regular, topological semigroup with identity $\underline{0}$ under the operation $+$ and topology $\mathcal{G}$.

0.1. Definition.

X is a uniform semigroup iff the topology $\mathcal{G}$ is induced by a uniformity $\mathcal{U}$ and the function $(x, y) \to x+y$ is uniformly continuous, namely, for every $U \in \mathcal{U}$ there exists a $V \in \mathcal{U}$ such that

$$(x, y) \in V \text{ and } (x', y') \in V \quad \Rightarrow \quad (x+x', y+y') \in U .$$

A key property of uniform semigroups is given by the following lemma.

Lemma. Let X be a uniform semigroup with uniformity $\mathcal{U}$. For every $U \in \mathcal{U}$ there exist $V_n \in \mathcal{U}$ for $n \in \omega$ such that, for every $N \in \omega$,

$$(x_n, y_n) \in V_n \text{ for } n = 0, \ldots, N \quad \Rightarrow \quad \left(\sum_{n=0}^{N} x_n, \sum_{n=0}^{N} y_n\right) \in U.$$

Proof: Let $V_{-1} = U$ and choose $V_n \in \mathcal{U}$ for $n \in \omega$, by recursion, so that

$$(x, y) \in V_n \text{ and } (x', y') \in V_n \Rightarrow (x+x', y+y') \in V_{n-1}. \blacksquare$$

Examples:

(1) Any topological group is a uniform semigroup.

(2) Let $X = [0, \infty]$ with the usual definition of $+$ and usual topology $\mathcal{G}$. Then X is compact and is a uniform semigroup. However, $(-\infty, \infty]$ is not a uniform semigroup even though its topology is induced by a uniformity.

(3) Let

E be a commutative topological group with identity 0,

$X = \{x : x \subset E \text{ and } x \neq \emptyset\}$

$x+y = \{e+e' ; e \in x \text{ and } e' \in y\}$ for $x,y \in X$.

Then X is a semigroup with identity $\underline{0} = \{0\}$. For each neighborhood U of 0 in E, let

$$U' = \{(x, y) : x,y \in X, \ x \subset y+U \text{ and } y \subset x+U\}$$

and $\mathcal{U}$ be the uniformity generated by the family of such U'. Under the topology $\mathcal{G}$ induced by $\mathcal{U}$, X is a uniform semigroup. The problem of integral representations for E-valued measures will lead us, in chapter III, to study the integral of X-valued functions.

0.2. Definition.

For any index set I , $x : I \to X$ and $y \in X$,

$\sum_{i \in I} x_i = y$ iff, for every neighborhood U of y , there exists a finite $J \subset I$ such that

$$J' \text{ finite and } J \subset J' \subset I \Rightarrow \sum_{i \in J'} x_i \in U .$$

0.3. Definitions.

For any $\tau : H \to X$,

(1) τ is finitely additive on H iff

$$\tau\Big(\bigcup_{\alpha \in F} \alpha\Big) = \sum_{\alpha \in F} \tau(\alpha)$$

for every non-empty, finite, disjoint $F \subset H$ with $\bigcup_{\alpha \in F} \alpha \in H$.

(2) τ is σ-additive on H iff

$$\tau\Big(\bigcup_{\alpha \in F} \alpha\Big) = \sum_{\alpha \in F} \tau(\alpha)$$

for every non-empty countable, disjoint $F \subset H$ with $\bigcup_{\alpha \in F} \alpha \in H$

0.4. Definitions.

(1) H is a *prering* iff, for every $A,B \in H$,

$$(A \cap B) \in H \text{ and } (A \setminus B) = \bigcup_{\alpha \in F} \alpha$$

for some finite, disjoint $F \subset H$.

(2) H is a *ring* iff, for every $A,B \in H$,

$$(A \cup B) \in H \text{ and } (A \setminus B) \in H .$$

(3) H is a *field* iff H is a ring and $S \in H$.

(4) H is a *σ-ring* iff, for every sequence A in H ,

$$\bigcup_{n \in \omega} A_n \in H \text{ and } (A_o \setminus A_1) \in H .$$

(5) H is a *σ-field* iff H is a σ-ring and $S \in H$.

(6) $\mathcal{B}(H)$ is the smallest σ-field containing H .

(7) $\text{sb}\, S = \{A : A \subset S\}$.

1. Approximation from above.

Let $\mu : \text{sb } S \to X$.

1.1. Definitions.

(1) μ is H-outer regular iff, for every $A \subset S$ and neighborhood U of $\mu(A)$, there exists $\alpha \in H$ such that $A \subset \alpha$ and, for every $\beta \in H$,

$$A \subset \beta \quad \Rightarrow \quad \mu(\alpha \cap \beta) \in H .$$

(2) α is a U-hull of A in (H, μ) iff $A \subset \alpha \in H$, $\mu(A) \in U \subset X$ and, for every E ,

$$A \subset E \subset \alpha \quad \Rightarrow \quad \mu(E) \in U .$$

(3) When X is a uniform semigroup with uniformity $\mathcal{U}$, for any $U \in \mathcal{U}$ we shall write " α is a U-hull of A " to mean " α is a V-hull of A " where $V = \{y : (\mu(A), y) \in U\}$.

1.2. Lemma. μ is H-outer regular iff, for every $A \subset S$ and neighborhood U of $\mu(A)$, there exists an α which is a U-hull of A in (H, μ) .

Proof: Suppose μ is H-outer regular. Given $A \subset S$ and neighborhood U of $\mu(A)$, let U' be a neighborhood of $\mu(A)$ with closure $U' \subset U$ and then choose $\alpha \in H$ such that $A \subset \alpha$ and, for every $\beta \in H$, $A \subset \beta \Rightarrow \mu(\alpha \cap \beta) \in U'$. Then, for any E with $A \subset E \subset \alpha$, we must have $\mu(E) \in$ closure U', for if V is any neighborhood of $\mu(E)$ there exists $\gamma \in H$ with $E \subset \gamma$ and $\mu(\beta \cap \gamma) \in V$ whenever $E \subset \beta \in H$. Thus, $\mu(\alpha \cap \gamma) \in U' \cap V$ so $U' \cap V \neq \emptyset$. ▮

2. <u>Measurable and Null Sets</u>.

Let $\mu : \text{sb}\, S \to X$.

2.1. <u>Definitions</u>.

(1) A is μ-<u>measurable</u> iff $A \subset S$ and, for every $T \subset S$,

$$\mu(T) = \mu(T \cap A) + \mu(T \setminus A) .$$

$$\mathcal{M}_\mu = \{A : A \text{ is } \mu\text{-measurable}\} .$$

(2) A is μ-<u>null</u> iff, for every $E \subset A$, $\mu(E) = \underline{0}$.

$$\mathcal{N}_\mu = \{A : A \text{ is } \mu\text{-null}\} .$$

2.2. <u>Lemma</u>.

If $\mathcal{M}_\mu \neq \emptyset$ then $\mathcal{M}_\mu$ is a field and μ is finitely additive on $\mathcal{M}_\mu$.

[Note that $\mu(\emptyset) = \underline{0}$ implies $\emptyset \in \mathcal{M}_\mu$.]

<u>Proof</u>: Clearly, $A \in \mathcal{M}_\mu$ implies $(S \setminus A) \in \mathcal{M}_\mu$. Given $A, B \in \mathcal{M}_\mu$, to check that $(A \cap B) \in \mathcal{M}_\mu$, let $T \subset S$. Then

$$\begin{aligned}\mu(T) &= \mu(T \cap A) + \mu(T \setminus A)\\ &= \mu((T \cap A) \cap B) + \mu((T \cap A) \setminus B) + \mu(T \setminus A) .\end{aligned}$$

But

$$\mu(T \setminus (A \cap B))$$
$$= \mu((T \setminus (A \cap B)) \cap A) + \mu(T \setminus (A \cap B) \setminus A)$$
$$= \mu((T \cap A) \setminus B) + \mu(T \setminus A) \quad .$$

Therefore,

$$\mu(T) = \mu(T \cap (A \cap B)) + \mu(T \setminus (A \cap B)) \ .$$

Thus, if $\mathcal{M}_\mu \neq \emptyset$ it is a field. It follows trivially from the definition that μ is finitely additive on $\mathcal{M}_\mu$ .I

2.3. Theorem. Let μ be H-outer regular. If μ is finitely additive on a ring $\mathcal{B}$ containing H then

(i) $\mathcal{B} \subset \mathcal{M}_\mu$ and

(ii) $\mathcal{N}_\mu \subset \mathcal{M}_\mu$ (hence $A,B \in \mathcal{N}_\mu \Rightarrow (A \cup B) \in \mathcal{N}_\mu$)

[Thus, $\mathcal{M}_\mu$ is the largest possible ring containing H on which μ is finitely additive and, moreover, $\mathcal{M}_\mu$ is complete relative to μ .]

Proof:

(i) Let $A \in \mathcal{B}$, $T \subset S$, U be any neighborhood of $\mu(T)$ and V be any neighborhood of $\mu(T \cap A) + \mu(T \setminus A)$. To see that

$U \cap V \neq \emptyset$, let V_1 be a neighborhood of $\mu(T \cap A)$ and V_2 a neighborhood of $\mu(T \setminus A)$ with $V_1 + V_2 \subset V$. Let

α be a U-hull of T in (H, μ) ,

α_1 be a V_1-hull of $(T \cap A)$ in (H, μ) ,

α_2 be a V_2-hull of $(T \setminus A)$ in (H, μ) ,

and set

$$E = (\alpha \cap \alpha_1 \cap A) \cup (\alpha \cap \alpha_2 \setminus A) .$$

Then $T \subset E \subset \alpha$ so $\mu(E) \in U$. On the other hand, since μ is finitely additive on $\mathcal{B}$, we have

$$\mu(E) = \mu(\alpha \cap \alpha_1 \cap A) + \mu(\alpha \cap \alpha_2 \setminus A)$$
$$\in V_1 + V_2 \subset V .$$

Thus, $U \cap V \neq \emptyset$ and $\mu(T) = \mu(T \cap A) + \mu(T \setminus A)$.

(ii) Let $A \in \mathcal{N}_\mu$ and $T \subset S$. Since $\mu(T \cap A) = \underline{0}$, we need only check that $\mu(T) = \mu(T \setminus A)$. Given any neighborhood U of $\mu(T)$ and any neighborhood V of $\mu(T \setminus A)$ we shall show that $U \cap V \neq \emptyset$. Let W be a closed neighborhood of $\underline{0}$ with

$$\mu(T) + W \subset U \qquad \text{and} \qquad \mu(T \setminus A) + W \subset V$$

and let α be a W-hull of A in (H, μ) . For any $E \subset \alpha$, we must have $\mu(E) \in W$ since

$$E \subset \beta \in H \quad \Rightarrow \quad A \subset (\beta \cap \alpha) \cup (A \setminus \beta) \subset \alpha$$

$$\Rightarrow \quad \mu(\beta \cap \alpha) = \mu(\beta \cap \alpha) + \mu(A \setminus \beta)$$

$$[\text{by (i)}] \quad = \mu((\beta \cap \alpha) \cup (A \setminus \beta)) \in W \ .$$

Now, by (i) ,

$$\mu(T) = \mu(T \cap \alpha) + \mu(T \setminus \alpha)$$

$$\mu(T \setminus A) = \mu((T \setminus A) \cap \alpha) + \mu((T \setminus A) \setminus \alpha)$$

$$= \mu((T \setminus A) \cap \alpha) + \mu(T \setminus \alpha) \ .$$

Hence

$$\mu(T) + \mu(T \setminus A \cap \alpha) = \mu(T \setminus A) + \mu(T \cap \alpha)$$

so

$$(\mu(T) + W) \cap (\mu(T \setminus A) + W) \neq \emptyset$$

therefore, $U \cap V \neq \emptyset$. I

3. Outer Measures.

Let $\mu : \text{sb } S \to X$.

3.1. Definitions.

(1) μ is an H-outer measure iff $\mu(\emptyset) = \underline{0}$, μ is σ-additive on $\mathcal{B}(H)$ and μ is H-outer regular.
(= smallest σ-field, containing H)

(2) μ is an outer measure iff μ is an H-outer measure for some H.

3.2. Theorem. Let μ be an H-outer measure. Then

(i) $A_n \subset A_{n+1} \subset S$ for $n \in \omega$ $\Rightarrow$ $\mu(\bigcup_{n\in\omega} A_n) = \lim_n \mu(A_n)$

(ii) $A_n \in \mathcal{N}_\mu$ for $n \in \omega$ $\Rightarrow$ $\bigcup_{n\in\omega} A_n \in \mathcal{N}_\mu$

(iii) $\mathcal{N}_\mu \cup \mathcal{B}(H) \subset \mathcal{M}_\mu$

(iv) $\mathcal{M}_\mu$ is a σ-field and μ is σ-additive on $\mathcal{M}_\mu$.

[Thus, $\mathcal{M}_\mu$ is the largest σ-field containing H on which μ is σ-additive. Moreover, $\mathcal{M}_\mu$ is complete relative to μ .]

Proof:

(i) Let $A_n \subset A_{n+1} \subset S$ and $B = \bigcup_{n\in\omega} A_n$. If $\mu(B) \neq \lim_n \mu(A_n)$ we we may suppose, without loss of generality, that there exists an open U with $\mu(B) \in U$ and $\mu(A_n) \in S \setminus \text{closure}\, U = V$ for every $n \in \omega$. Let β be a U-hull of B, α_n be a V-hull of A_n and set

$$\gamma_n = \bigcap_{i \geq n} (\beta \cap \alpha_i) .$$

Then

$$A_n \subset \gamma_n \subset \alpha_n \quad \text{and} \quad B \subset \bigcup_{n \in \omega} \gamma_n \subset \beta$$

so

$$\mu(\gamma_n) \in V \quad \text{and} \quad \mu(\bigcup_{n \in \omega} \gamma_n) \in U$$

and, since $\gamma_n \subset \gamma_{n+1} \in \mathcal{B}(H)$ and μ is σ-additive on $\mathcal{B}(H)$, we have

$$\mu(\bigcup_{n \in \omega} \gamma_n) = \lim_n \mu(\gamma_n) .$$

Since U is open, this implies that, for large n, $\mu(\gamma_n) \in U$ and therefore $U \cap V \neq \emptyset$ which is impossible.

(ii) Immediate from 2.3(ii) and (i) above.

(iii) Immediate from 2.3.

(iv) Let

$$A_n \subset A_{n+1} \in \mathcal{M}_\mu \quad \text{and} \quad B = \bigcup_{n \in \omega} A_n .$$

Then, for any $T \subset S$,

$$T = \bigcup_{n \in \omega} ((T \setminus B) \cup (T \cap A_n))$$

therefore,

$$\begin{aligned}
\mu(T) &= \lim_n \mu((T \setminus B) \cup (T \cap A_n)) && \text{by (i)} \\
&= \lim_n \; [\mu(T \setminus B) + \mu(T \cap A_n)] && \text{since } A_n \in \mathcal{M}_\mu \\
&= \mu(T \setminus B) + \lim_n \mu(T \cap A_n) \\
&= \mu(T \setminus B) + \mu(T \cap B) \quad . && \text{by (i)}
\end{aligned}$$

Thus, $B \in \mathcal{M}_\mu$ and, by 2.2, $\mathcal{M}_\mu$ is a σ-field. By 2.2 and (i) above, μ is σ-additive on $\mathcal{M}_\mu$. ∎

Exercise.

Let $X = [0, \infty]$ with the usual operation $+$ and topology $\mathcal{G}$. Then μ is an outer measure iff $\mu(\emptyset) = 0$ and, for every $A \subset S$,

(i) $A \subset \bigcup_{n \in \omega} B_n \subset S \quad \Rightarrow \quad \mu(A) \leq \sum_{n \in \omega} \mu(B_n)$

(ii) $\mu(A) = \inf\{\mu(\alpha) \; ; \; \alpha \in \mathcal{M}_\mu \text{ and } A \subset \alpha\}$

namely, μ is an outer measure in the sense of Caratheodory.

4. Generation of an Outer Measure: the Integral Process.

It is very seldom that one is handed a ready made σ-additive function on a σ-field. As most such functions are manufactured from more primitive material, the generation of a measure is one of the fundamental problems in measure theory. Most frequently, it appears in the guise of extending a σ-additive function on a ring to one on a σ-field. In functional analysis, it takes the form of extending the notion of integral to a wider class of functions or, more generally, of extending some linear functional.

When $X = [0, \infty]$, the Caratheodory process for generating an outer measure is one of the most powerful tools available. From any $\tau : H \to X$ it produces a function μ through the formula

$$\mu(A) = \inf\{ \sum_{\alpha \in F} \tau(\alpha) \ ; \ F \text{ a countable subfamily of } H \text{ which covers } A\}$$

for any $A \subset S$.

Another process, used when S is a topological space and H is the family of compact sets, is to define μ through the formula

$$\mu(A) = \inf_{\substack{U \text{ open} \\ A \subset U}} \ \sup_{\substack{\alpha \in H \\ \alpha \subset U}} \tau(\alpha) \quad .$$

A third major process is to define first the integral $\int f d\tau$ as the limit of sums and then set

$$\mu(A) = \int 1_A \, d\tau \quad .$$

We shall combine the ideas in all three processes mentioned above and define directly, for a given $\tau : H \to X$ and any $A \subset S$,

$$\int_A d\tau .$$

This will solve simultaneously the problem of generating an outer measure μ and of producing an integral, for as we shall see in the next chapter the seemingly more general concept of $\int f\cdot d\tau$ will be a special case of the above.

4.1. Definition of the Integral Process.

Given $\tau : H \to X$, for any $A \subset S$, let

(1) $\mathcal{P}(A) = \{P : P$ is a countable, disjoint subfamily of H which covers $A\}$.

(2) for $P, Q \in \mathcal{P}(A)$,

$P < Q$ iff Q is a refinement of P, i.e., for every $\beta \in Q$ there exists an $\alpha \in P$ with $\beta \subset \alpha$.

(3) $\mathscr{P}(A)$ be the set of all pairs (P, Δ) such that

(i) $P \in \mathcal{P}(A)$ and

(ii) Δ is a function on $\{Q \in \mathcal{P}(A) : P < Q\}$ such that $\Delta(Q)$ is finite and $\Delta(Q) \subset Q$ for every Q in its domain.

(4) for $(P, \Delta) \in \mathcal{D}(A)$ and $(P', \Delta') \in \mathcal{D}(A)$,

$(P, \Delta) \ll (P', \Delta')$ iff $P < P'$ and $\Delta(Q) \subset \Delta'(Q)$ for every $Q \in \mathcal{P}(A)$ with $P' < Q$.

(5) $\int_A d\tau = \text{limit} \sum_{\alpha \in \Delta(P)} \tau(\alpha)$ as (P, Δ) runs over $(\mathcal{D}(A), \ll)$

Note that $\mathcal{P}(A)$ may not be directed by $<$, and hence $\mathcal{D}(A)$ may not be directed by $\ll$, unless we impose some conditions on H. Also, the limit in (5) may be very different from

$$\text{limit} \sum_{\alpha \in P} \tau(\alpha) \text{ as } P \text{ runs over } (\mathcal{P}(A), <) .$$

For example, $\sum_{\alpha \in P} \tau(\alpha)$ may fail to exist for sufficiently fine P and yet $\int_A d\tau$ may exist. The key idea of R. S. Phillips was to replace the limit of infinite sums (two iterated limits) by a (double) limit of finite sums.

Finally, note that when τ is σ-additive on H we have $\tau(\alpha) = \int_\alpha d\tau$ for every $\alpha \in H$.

The major properties of $\int_A d\tau$ are summarized in the following theorems.

4.2. Theorem. Let $\tau : H \to X$ and suppose, for every $A \subset S$, $\mathcal{P}(A)$ is directed by $<$ and $\mu(A) = \int_A d\tau \in X$. Then

(1) μ is H_σ-outer regular;

(2) for every non-empty, finite, disjoint $F \subset H$,

$$\mu(\bigcup_{\alpha \in F} \alpha) = \sum_{\alpha \in F} \mu(\alpha) \quad ;$$

(3) when X is a uniform semigroup, for every non-empty, countable, disjoint $F \subset H$,

$$\mu(\bigcup_{\alpha \in F} \alpha) = \sum_{\alpha \in F} \mu(\alpha) \quad .$$

Proof:

(1) Given $A \subset S$ and a closed neighborhood U of $\mu(A)$, choose $(P, \Delta) \in \mathcal{P}(A)$ so that, for every $(Q, \eta) \in \mathcal{P}(A)$,

$$(P, \Delta) << (Q, \eta) \quad \Rightarrow \quad \sum_{\beta \in \eta(Q)} \tau(\beta) \in U$$

and let

$$\alpha = \bigcup_{\beta \in P} \beta \quad .$$

Then $A \subset \alpha \in H_\sigma$. To see that α is a U-hull of A, let $A \subset E \subset \alpha$ and V be any neighborhood of $\mu(E)$. We shall show that $U \cap V \neq \emptyset$. To this end, choose $(P_1, \Delta_1) \in \mathcal{P}(E)$ so that, for every $(Q, \eta) \in \mathcal{P}(E)$,

$$(P_1, \Delta_1) << (Q, \eta) \quad \Rightarrow \quad \sum_{\beta\in\eta(Q)} \tau(\beta) \in V .$$

Next, choose $P' \in \mathcal{P}(E)$ so that $P < P'$ and $P_1 < P'$ and let

$$\Delta'(Q) = \begin{cases} \Delta(Q) \cup \Delta_1(Q) & \text{for } Q \in \mathcal{P}(E),\ P' < Q \\ \Delta(Q) & \text{for } Q \in \mathcal{P}(A) \setminus \mathcal{P}(E) . \end{cases}$$

Since $\mathcal{P}(E) \subset \mathcal{P}(A)$, we have $(P', \Delta') \in \mathcal{D}(A) \cap \mathcal{D}(E)$,

$$(P, \Delta) << (P', \Delta') \quad \text{in } \mathcal{D}(A)$$

$$(P_1, \Delta_1) << (P', \Delta') \quad \text{in } \mathcal{D}(E)$$

hence

$$\sum_{\beta\in\Delta'(P')} \tau(\beta) \in U \cap V .$$

Thus, $U \cap V \neq \emptyset$ and $\mu(E) \in$ closure $U = U$.

(2) Let F be a non-empty, finite, disjoint subfamily of H and $A = \bigcup_{\alpha\in F} \alpha$. Given a neighborhood U of $\mu(A)$ and a neighborhood V of $\sum_{\alpha\in F} \mu(\alpha)$, we shall show $U \cap V \neq \emptyset$. First, choose $(P, \Delta) \in \mathcal{D}(A)$ so that $F < P$ and, for every $(Q, \eta) \in \mathcal{D}(A)$,

$$(P, \Delta) << (Q, \eta) \quad \Rightarrow \quad \sum_{\beta\in\eta(Q)} \tau(\beta) \in U .$$

Next, for each $\alpha \in F$, choose a neighborhood V_α of $\mu(\alpha)$ so that

$$\sum_{\alpha\in F} V_\alpha \subset V$$

and $(P_\alpha, \Delta_\alpha) \in \mathcal{D}(\alpha)$ so that $\{\beta \in P : \beta \subset \alpha\} < P_\alpha$ and, for every $(Q, \eta) \in \mathcal{D}(\alpha)$,

$$(P_\alpha, \Delta_\alpha) << (Q, \eta) \quad \Rightarrow \quad \sum_{\beta \in \eta(Q)} \tau(\beta) \in V_\alpha .$$

Let

$$P' = \bigcup_{\alpha \in F} P_\alpha$$

$$\Delta'(P') = \Delta(P') \cup \bigcup_{\alpha \in F} \Delta_\alpha(P_\alpha)$$

and $\Delta'(Q) = \Delta(Q)$ for $Q \in \mathcal{P}(A)$, $P' < Q$, $P' \neq Q$. Then $(P, \Delta) << (P', \Delta')$ so

$$\sum_{\beta \in \Delta'(P')} \tau(\beta) \in U .$$

On the other hand, for each $\alpha \in F$, let

$$\eta_\alpha(P_\alpha) = \Delta'(P') \cap P_\alpha \qquad [\supset \Delta_\alpha(P_\alpha)]$$

and $\eta_\alpha(Q) = \Delta_\alpha(Q)$ for $Q \in \mathcal{P}(\alpha)$, $P_\alpha < Q$, $P_\alpha \neq Q$. Then $(P_\alpha, \Delta_\alpha) << (P_\alpha, \eta_\alpha)$ and so

$$\sum_{\beta \in \Delta'(P')} \tau(\beta) = \sum_{\alpha \in F} \sum_{\beta \in \eta_\alpha(P_\alpha)} \tau(\beta) \in \sum_{\alpha \in F} V_\alpha \subset V .$$

Thus, $U \cap V \neq \emptyset$.

(3) Let $\mathcal{U}$ be a uniformity for X, $\alpha_n \in H$ with $\alpha_n \cap \alpha_m = \emptyset$ for $n,m \in \omega$, $n \neq m$, and $A = \bigcup_{n \in \omega} \alpha_n$. Given $U \in \mathcal{U}$, we

shall determine a finite $K \subset \omega$ so that for every finite $N \subset \omega$,

$$K \subset N \quad \Rightarrow \quad (\mu(A), \sum_{n \in N} \mu(\alpha_n)) \in U .$$

To this end, choose $U' \in \mathcal{U}$ and $V_n \in \mathcal{U}$ so that $U' \circ U' \subset U$ and, for every $N \subset \omega$,

$$(x_n, y_n) \in V_n \text{ for } n \in N \quad \Rightarrow \quad (\sum_{n \in N} x_n, \sum_{n \in N} y_n) \in U' .$$

Next, choose $(P, \Delta) \in \mathfrak{D}(A)$ so that $\{\alpha_n ; n \in \omega\} < P$ and, for every $(Q, \eta) \in \mathfrak{D}(A)$,

$$(P, \Delta) << (Q, \eta) \quad \Rightarrow \quad (\mu(A), \sum_{\beta \in \eta(Q)} \tau(\beta)) \in U'$$

Finally, choose $(P_n, \Delta_n) \in \mathfrak{D}(\alpha_n)$ so that

$\{\beta \in P : \beta \subset \alpha_n\} < P_n$ and, for every $(Q, \eta) \in \mathfrak{D}(\alpha_n)$,

$$(P_n, \Delta_n) << (Q, \eta) \quad \Rightarrow \quad (\sum_{\beta \in \eta(Q)} \tau(\beta), \mu(\alpha_n)) \in V_n .$$

Let

$$P' = \bigcup_{n \in \omega} P_n$$

and since $\Delta(P')$ is finite, let

$$K = \{n : \Delta(P') \cap P_n \neq \emptyset\} .$$

Given a finite $N \subset \omega$ with $K \subset N$, let

$$\Delta'(P') = \Delta(P') \cup \bigcup_{n \in N} \Delta_n(P_n)$$

and $\Delta'(Q) = \Delta(Q)$ for $Q \in \mathcal{P}(A)$, $Q \neq P'$, $P' < Q$. Then

$(P, \Delta) << (P', \Delta')$ so

$$(\mu(A), \sum_{\beta\in\Delta'(P')} \tau(\beta)) \in U' .$$

On the other hand, for $n \in N$, let

$$\eta_n(P_n) = \Delta'(P') \cap P_n$$

and $\eta_n(Q) = \Delta_n(Q)$ for $Q \in \mathcal{P}(\alpha_n)$, $Q \neq P_n$, $P_n < Q$. Then $(P_n, \Delta_n) << (P_n, \eta_n)$ so

$$(\sum_{\beta\in\eta_n(P_n)} \tau(\beta), \mu(\alpha_n)) \in V_n .$$

Also,

$$\sum_{\beta\in\Delta'(P')} \tau(\beta) = \sum_{n\in N} \sum_{\beta\in\eta_n(P_n)} \tau(\beta) .$$

Therefore

$$(\mu(A), \sum_{n\in N} \mu(\alpha_n)) \in U' \circ U' \subset U . \qquad \text{I}$$

4.3. <u>Theorem</u>. Let X be a uniform semigroup, H be a ring, and $\tau : H \to X$ with $\tau(\emptyset) = \underline{0}$. If, for every $A \subset S$, $\mu(A) = \int_A d\tau \in X$ then μ is an H_σ-outer measure.

Proof:

(1) For every countable $F \subset H$, $\mu(\bigcup_{\alpha \in F} \alpha) = \lim \mu(\bigcup_{\alpha \in J} \alpha)$ as J runs over the finite subsets of F directed by $\subset$.

Otherwise, there exists a neighborhood U of $\mu(\bigcup_{\alpha \in F} \alpha)$ and finite $J_n \subset F$ for $n \in \omega$ such that $J_n \subset J_{n+1}$, $\bigcup_{n \in \omega} J_n = F$ and $\mu(\bigcup_{\alpha \in J_n} \alpha) \notin U$. Letting $\beta_{-1} = \emptyset$ and

$$\beta_n = \bigcup_{\alpha \in J_n} \alpha \setminus \beta_{n-1}$$

we get $\beta_n \in H$, $\beta_n \cap \beta_m = \emptyset$ for $n \neq m$ and

$$\sum_{n \in \omega} \mu(\beta_n) \neq \mu(\bigcup_{n \in \omega} \beta_n) = \mu(\bigcup_{\alpha \in F} \alpha)$$

in contradiction to 4.2(3).

(2) $H_\sigma \subset \mathfrak{M}_\mu$.

Given $A \in H_\sigma$, $T \subset S$, any open neighborhood U of $\mu(A)$ and neighborhood V of $\mu(T \cap A) + \mu(T \setminus A)$ we shall show $U \cap V \neq \emptyset$. Let V_1 be an open neighborhood of $\mu(T \cap A)$ and V_2 be an open neighborhood of $\mu(T \setminus A)$ with $V_1 + V_2 \subset V$. By 4.2(1), let $T', T_1, T_2 \in H_\sigma$ be such that

T' is a U-hull of A,

T_1 is a V_1-hull of $T \cap A$, $T_1 \subset T' \cap A$,

T_2 is a V_2-hull of $T \setminus A$, $T_2 \subset T'$.

Let

$$T_1 = \bigcup_{m\in\omega} \alpha_m \qquad \text{with } \alpha_m \subset \alpha_{m+1} \in H$$

$$T_2 = \bigcup_{n\in\omega} \beta_n \qquad \text{with } \beta_n \subset \beta_{n+1} \in H \quad .$$

Then

$$T \subset \bigcup_{n\in\omega} \bigcup_{m\in\omega} (\alpha_m \cup \beta_n) \subset T'$$

hence, by (1), there exists $N \in \omega$ such that, for $m,n \in \omega$, $m > N$ and $n > N \Rightarrow \mu(\alpha_m \cup \beta_n) \in U$. Choose, first $m \in \omega$ so that $m > N$ and $\mu(\alpha_m) \in V_1$. Since

$$T \setminus A \subset T_2 \setminus \alpha_m = \bigcup_{n\in\omega} (\beta_n \setminus \alpha_m) \subset T_2 \quad ,$$

there exists $n > \omega$ such that $n > N$ and $\mu(\beta_n \setminus \alpha_m) \in V_2$. Thus, $\mu(\alpha_m \cup \beta_n) \in U$ and, by 4.2(2),

$$\mu(\alpha_m \cup \beta_n) = \mu(\alpha_m) + \mu(\beta_n \setminus \alpha_m) \in V_1 + V_2 \subset V$$

so $U \cap V \neq \emptyset$.

(3) $A_n \subset A_{n+1} \subset S$ for $n \in \omega \Rightarrow \mu(\bigcup_{n\in\omega} A_n) = \lim_n \mu(A_n)$.

Let $\mathcal{U}$ be the uniformity on X . Given $U \in \mathcal{U}$, choose $U' \in \mathcal{U}$ and $V_n \in \mathcal{U}$ so that $U' \circ U' \circ U' \subset U$ and, for every $N \in \omega$,

$$(x_n, y_n) \in V_n \text{ for } n = 0,\cdots,N \Rightarrow (\sum_{n=0}^{N} x_n, \sum_{n=0}^{N} y_n) \in U' \quad .$$

Given

$$A_n \subset A_{n+1} \subset S \quad \text{for} \quad n \in \omega \quad \text{and} \quad B = \bigcup_{n\in\omega} A_n \ ,$$

let $\beta \in H_\sigma$ be a U'-hull of B and then, by recursion, choose $\alpha_n \in H_\sigma$ for $n \in \omega$ such that $\alpha_n \subset \beta$, α_o is a V_o-hull of A_o , and

$$\alpha_{n+1} \text{ is a } V_{n+1}\text{-hull of } (A_{n+1} \setminus \bigcup_{i=0}^{n} \alpha_i) \ .$$

Then

$$B \subset \bigcup_{n\in\omega} \alpha_n \subset \beta$$

so, by (1), there exists $K \in \omega$ so that, for every $N \in \omega$,

$$K < N \quad \Rightarrow \quad (\mu(B), \mu(\bigcup_{n=0}^{N} \alpha_n)) \in U' \circ U' \ .$$

Let

$$\gamma_o = \alpha_o \quad \text{and} \quad \gamma_{n+1} = \alpha_n \setminus \bigcup_{i=0}^{n} \alpha_i \ ,$$

so

$$A_{n+1} \subset \gamma_{n+1} \subset \alpha_{n+1}$$

and

$$(\mu(\gamma_n), \mu(A_n)) \in V_n \ .$$

Therefore, for any $N \in \omega$, $N > K$, we have

$$\mu(\bigcup_{n=0}^{N} \alpha_n) = \sum_{n=0}^{N} \mu(\gamma_n)$$

hence

$$(\mu(B), \sum_{n=0}^{N} \mu(A_n)) \in U' \circ U' \circ U' \subset U \quad .$$

(4) $A_n \subset A_{n+1} \in \mathcal{M}_\mu$ for $n \in \omega$ $\Rightarrow$ $\bigcup_{n \in \omega} A_n \in \mathcal{M}_\mu$.

This is immediate from (3).

(5) $\mathcal{M}_\mu$ is a σ-field and μ is σ-additive on $\mathcal{M}_\mu$.

This is immediate from 2.2(3), (4) and 4.2(2) .

In view of 4.2(1), this completes the proof of the theorem.

I

4.4. Theorem. Let X be a uniform semigroup, $\tau : H \to X$, and suppose, for every $A \subset S$, $\mathcal{P}(A)$ is directed by $<$ and $\mu(A) = \int_A d\tau \in X$. Then

(1) $\mu(A) = \int_A d(\mu|H)$ for every $A \subset S$

(2) When H is a ring and $\tau(\emptyset) = \underline{0}$,

$\mu(A) = \int_A d(\mu|\mathcal{M}_\mu)$ for every $A \subset S$

Proof: (1) follows immediately from 4.2(1) and 4.2(3)

(2) follows immediately from 4.3 since, in this case, $H \subset \mathcal{M}_{\mu}$.

5. Existence Theorems.

We shall now study some special conditions under which $\int_A d\tau$ exists. Other conditions will be considered in the next chapter.

5.1. Definitions.

(1) For $\tau : H \to X$,

τ is s-bounded iff, for every disjoint sequence α in H, $\lim_n \tau(\alpha_n) = \underline{0}$.

(2) For any I directed by $<$ and $x : I \to X$,

x is an s-Cauchy net iff, for every neighborhood U of $\underline{0}$, there exists $i \in I$ such that, for every $j,k \in I$ with $i < j$ and $i < k$, we have

$$(x_j + U) \cap (x_k + U) \neq \emptyset .$$

(3) For $A \subset X$,

A is s-complete iff every s-Cauchy net in A converges to some point in A.

A is s-precomplete iff closure A is s-complete.

5.2. Theorem. Let H be a ring with $S \in H_\sigma$ and $\tau : H \to X$. If τ is finitely additive and s-bounded and range τ is s-precomplete then $\int_A d\tau \in X$ for every $A \subset S$.

Proof: Since τ is s-bounded, we have

(1) For any countable, disjoint $F \subset H$ and any neighborhood U of $\underline{0}$, there exists a finite $J \subset F$ such that, for every finite $J' \subset F$ and $\beta \in H$,

$$J' \cap J = \emptyset \text{ and } \beta \subset \bigcup_{\alpha \in J'} \alpha \quad \Rightarrow \quad \tau(\beta) \in U .$$

Since τ is finitely additive and range τ is s-precomplete, we conclude

(2) For any countable, disjoint $F \subset H$, $\sum_{\alpha \in F} \tau(\alpha) \in X$.

Let $A \subset S$ and, for any $P \in \mathcal{P}(A)$, let

$$t(P) = \sum_{\alpha \in P} \tau(\alpha) .$$

If $\int_A d\tau \notin X$ then there exist a neighborhood U of $\underline{0}$ and $P_n \in \mathcal{P}(A)$ for $n \in \omega$ such that $P_n < P_{n+1}$ but

$$(t(P_n) + U) \cap (t(P_{n+1}) + U) = \emptyset .$$

For $n \in \omega$, let V_n be a closed neighborhood of $\underline{0}$ such that $V_o + V_o + V_o \subset U$ and $V_{n+1} + V_{n+1} \subset V_n$. In view of (1), choose finite $J_n \subset P_n$ so that, for every finite $J' \subset P_n$ and $\beta \in H$,

$$J' \cap J_n = \emptyset \text{ and } \beta \subset \bigcup_{\alpha \in J'} \alpha \quad \Rightarrow \quad \tau(\beta) \in V_n .$$

Let

$$\beta_n = \bigcup_{\alpha \varepsilon J_n} \alpha$$

$$\gamma_n = \bigcap_{i=0}^{n} \beta_i \quad .$$

Then $\gamma_{n+1} \subset \gamma_n \varepsilon H$ so $\lim_n \tau(\gamma_n \setminus \gamma_{n+1}) = \underline{0}$ hence, for some $N \varepsilon \omega$,

$$\tau(\gamma_N \setminus \gamma_{N+1}) \varepsilon V_o \qquad \text{and} \qquad \tau(\gamma_N) \varepsilon \tau(\gamma_{N+1}) + V_o \quad .$$

Now,

$$\begin{aligned} \beta_n \setminus \gamma_n &= \bigcup_{i=0}^{n} (\beta_n \setminus \beta_i) \\ &= (\beta_n \setminus \beta_o) \cup (\beta_n \cap \beta_o \setminus \beta_1) \\ &\quad \cup (\beta_n \cap \beta_0 \cap \beta_1 \setminus \beta_2) \cup \cdots\cdot \end{aligned}$$

and, for $i = 0, \cdots, n$,

$$\beta_n \setminus \beta_i \subset \bigcup_{\alpha \varepsilon J'} \alpha$$

for some finite $J' \subset P_i$ with $J' \cap J_i = \emptyset$. Hence

$$\tau(\beta_n \setminus \gamma_n) \varepsilon \sum_{i=0}^{n-1} V_i$$

and $$t(P_n) \in \tau(\beta_n) + V_n \subset \tau(\gamma_n) + \sum_{i=0}^{n} V_i$$

$$\subset \tau(\gamma_n) + V_o + V_o \quad .$$

Thus,

$$t(P_N) \in \tau(\gamma_N) + V_o + V_o \subset \tau(\gamma_{N+1}) + V_o + V_o + V_o$$

$$\subset \tau(\gamma_{N+1}) + U$$

and $$t(P_{N+1}) \in \tau(\gamma_{N+1}) + V_o + V_o \subset \tau(\gamma_{N+1}) + U$$

so

$$(t(P_N) + U) \cap (t(P_{N+1}) + U) \neq \emptyset$$

[since $x = a + u$ and $y = a + u'$ $\Rightarrow$ $x + u' = y + u$] .

This contradicts the assumption about the P_n . I

5.3. <u>Theorem.</u> Let X be a complete group, $\tau : H \to X$ with $\tau(\emptyset) = \underline{0}$, and suppose $\int_S d\tau \in X$.

(1) If H is a ring then, for every $\alpha \in H$, $\int_\alpha d\tau \in X$.

(2) If H is a σ-ring then, for every $A \subset S$, $\int_A d\tau \in X$.

Proof:

(1) Let H be a ring and $\alpha \in H$. We shall show that, as (P_1, Δ_1) runs over $(\mathfrak{D}(\alpha), <<)$, the sums

$$\sum_{\beta \in \Delta_1(P_1)} \tau(\beta)$$

form a Cauchy net. Given a neighborhood U of $\underline{0}$, choose $(P, \Delta) \in \mathfrak{D}(S)$ so that, for every $(Q, \eta) \in \mathfrak{D}(S)$,

$$(P, \Delta) << (Q, \eta) \quad \Rightarrow \quad \sum_{\beta \in \eta(Q)} \tau(\beta) - \int_S d\tau \in U .$$

Let

$$P_1 = \{\beta \cap \alpha \ ; \ \beta \in P\} \ ,$$

$$P_2 = \{\beta \setminus \alpha \ ; \ \beta \in P\} \ ,$$

$$P' = P_1 \cup P_2 \ ,$$

and, for any $Q_1 \in \mathfrak{P}(\alpha)$ with $P_1 < Q_1$, let

$$Q = Q_1 \cup P_2$$

$$\Delta_1(Q_1) = \Delta(Q) \cap Q_1 \ .$$

Then, for any $(Q_1, \eta_1) \in \mathfrak{D}(\alpha)$ with $(P_1, \Delta_1) << (Q_1, \eta_1)$, letting

$$\eta(Q) = \eta_1(Q_1) \cup (\Delta(Q) \cap P_2) \cup (\Delta(P') \cap P_2) \ ,$$

$$\Delta'(P') = \Delta(P') \cup (\Delta(Q) \cap P_2) \ ,$$

and $\eta(Q') = \Delta'(Q') = \Delta(Q')$ for any $Q' \in \mathcal{P}(S)$ with $P' < Q'$, $P' \neq Q'$, $Q \neq Q'$, we see that

$$(P, \Delta) << (P', \Delta') \quad \text{and} \quad (P, \Delta) << (Q, \eta)$$

hence

$$\sum_{\beta\in\eta(Q)} \tau(\beta) - \sum_{\beta\in\Delta'(P')} \tau(\beta) \in U + U ,$$

and therefore

$$\sum_{\beta\in\eta_1(Q_1)} \tau(\beta) - \sum_{\beta\in\Delta_1(P_1)} \tau(\beta) \in U + U .$$

(2) Let H be a σ-ring and $A \subset S$. For $\alpha \in H$, let $\mu(\alpha) = \int_\alpha d\tau$. In view of (1) above and 4.2(3), in the proof of which only the hypothesis $\mu(A) \in X$ for $A \in H_\sigma$ was used, we conclude that for any descending sequence α in H , $\lim_n \mu(\alpha_n) \in X$. If $\int_A d\tau \notin X$ then, as α runs over $\{\alpha \in H : A \subset \alpha\}$, the $\mu(\alpha)$ would not form a Cauchy net and hence there would exist a descending sequence α in H for which $\lim_n \mu(\alpha_n) \notin X$. I

6. Extension Theorems.

Putting together the results of sections 4 and 5, we get the following.

6.1. Theorem. Let X be a uniform semigroup, H be a ring with $S \in H_\sigma$, $\tau : H \to X$ and

$$\mu(A) = \int_A d\tau \qquad \text{for } A \subset S .$$

If τ is finitely additive and s-bounded and range τ is s-precomplete then μ is an H_σ-outer measure. If, in addition, τ is σ-additive then μ is an extension of τ.

Proof: Apply theorems 5.2 and 4.3. ∎

Remark. Note that if a subset of a uniform semigroup is complete then it is s-complete.

6.2. Theorem. Let X be a group, H be a ring with $S \in H_\sigma$, $\tau : H \to X$ and range τ be contained in a complete subset of X. Then τ can be extended to an H_σ-outer measure iff τ is σ-additive on H and,

(*) for every increasing sequence α in H, $\lim_n \tau(\alpha_n) \in X$.

Proof: The necessity is trivial. Suppose now that τ is σ-additive and condition (*) holds. Since X is a group, this implies that τ is s-bounded. Moreover, in a group, s-completeness is equivalent to completeness. Therefore, letting

$$\mu(A) = \int_A d\tau$$

for every $A \subset S$, we see by 6.1 that μ is an H_σ-outer measure which extends τ. ∎

6.3. Remark. We should note that the extension μ obtained above is unique. In fact, it is characterized by the formulas:

(1) for $A \in H_\sigma$, $\mu(A) = \text{limit } \tau(\alpha)$ as α runs over $\{\alpha \in H : \alpha \subset A\}$ directed by $\subset$;

(2) for $E \subset S$, $\mu(E) = \text{limit } \mu(A)$ as A runs over $\{A \in H_\sigma : E \subset A\}$ directed by $\supset$.

This is immediate from the properties of τ and the definition of $\int_E d\tau$, in view of theorems 5.2 and 4.3.

REFERENCES

1. Phillips, R. S. Integration in a convex linear topological space. Trans. A.M.S., 47(1940), pp. 114-145.

2. Rickart, C. E. Integration in a convex linear topological space. Trans. A.M.S., 52(1942), pp. 498-521.

3. ____________, Decomposition of additive set functions. Duke Math. J., 10(1943), pp. 653-665.

4. Sion, M. Introduction to the methods of real analysis. Holt, Rinehart and Winston, New York 1968.

5. ____________, Outer measures with values in a topological group. Proc. London Math. Soc., 19(1969), pp. 89-106.

6. ____________, Lectures on vector valued measures. University of British Columbia, 1969-70.

7. ____________, Group-valued outer measures. Actes du Congres International des Mathematiciens, Nice 1970, vol. 2, pp. 589-593.

8. Traynor, T. Differentiation of group-valued outer measures. Thesis, University of British Columbia, 1969.

9. ____________, A general Hewitt-Yosida decomposition. (To appear)

The concept of a group-valued outer measure first appeared in Sion [5] where the problem of extending a group-valued σ-additive function on a ring to one on a σ-ring (theorem 6.2) was solved. There, the outer measure was generated through the formulas given in remark 6.3.

The integral process in 4.1 is a refinement of the definitions of integral given by Phillips [1], Rickart [2], and Traynor [8].

The idea of using the integral process to generate the outer measure is strongly hinted at in Sion [4] and explicitely used in Sion [6].

The theorems in sections 4 and 5 are based on the work in Sion [5, 6]. Theorem 5.2 in its present form, however, is a refinement of a result in Traynor [9].

The notion of s-boundedness is found in Rickart [3]. The notion of s-completeness is new.

When S is a topological space, the notion of a Radón outer measure is considered in Sion [5].

CHAPTER II

MEASURABLE AND INTEGRABLE FUNCTIONS

We shall now study $\int_A d\tau$ in the special case when $\tau(\alpha)$ is of the form $f(\alpha)\cdot\lambda(\alpha)$. This leads to the notion of the integral of f with respect to λ , denoted by $\int_A f\cdot d\lambda$. Although, a priori, the roles played by f and λ should be symmetric, we shall follow by and large the traditional view of fixing the 'measure' λ in our mind and of studying properties of the 'function' f in relation to λ . In order to offer maximum flexibility for applications, the ranges of f , λ , and τ will be kept separate and hence the 'multiplication' operation '$\cdot$' need not be commutative. The use of this binary operation is purely a notational and psychological convenience rather than a logical necessity.

In the above generality, the concept of integral is really equivalent to that of the previous chapter, for the latter corresponds to a choice of f and $\cdot$ which yields $f(\alpha)\cdot\lambda(\alpha) = \lambda(\alpha)$. In view of the applications we have in mind, especially the integral of a derivative discussed in the next chapter, we shall gradually specialize f , $\cdot$, and λ quite a bit.

1. Integral of a Function.

We suppose

S , X_1 , X_2 are abstract spaces;

H is a family of subsets of S ;

X is a commutative, topological semigroup;

$\cdot : (a, b) \in X_1 \times X_2 \to a \cdot b \in X$;

$\lambda : H \to X_2$.

We are interested in integrating a 'multiple-valued' function f on S to X_1 with respect to λ and obtaining a single point in X as an answer.

1.1. Definitions.

(1) f is a multiple-valued function on S to X_1 iff

$$f : S \to \{E : E \subset X_1 \text{ and } E \neq \emptyset\} .$$

Such an f is single-valued iff, for every $s \in S$, $f(s)$ is a singleton.

(2) For any multiple-valued function f on S to X_1 ,

$$f[A] = \bigcup_{s \in A} f(s) \qquad \text{for } A \subset S$$

$$f^{-1}[E] = \{s \in S : f(s) \subset E\} \qquad \text{for } E \subset X_1 .$$

(3) We shall consider $f : S \to X_1$ as a single-valued function on S to X_1 so $f[A]$ is the image of A under f and $f^{-1}[E]$ is the inverse image of E under f .

1.2. Definitions.

For any multiple-valued function f on S to X_1 ,

(1) ξ is a choice function for f iff, for every $\alpha \in H$,

$$\xi(\alpha) \in f[\alpha] \quad .$$

(2) for any such ξ and $\alpha \in H$,

$$\tau_\xi(\alpha) = \xi(\alpha)\cdot\lambda(\alpha)$$

(3) for any $A \subset S$,

$\int_A f\cdot d\lambda$ = The $x \in X$ (when it exists) such that ,

for every choice function ξ for f ,

$$x = \int_A d\tau_\xi \quad .$$

The above definitions have been formulated in a way which avoids consideration of spaces and operations other than the given ones and emphasizes our interest in obtaining the integral as a point in X even though the integrand is permitted to be multiple-valued.

For the sake of perspective, however, it should be pointed out that, even when f is single-valued, one is naturally led to consider the integral as a subset of X. The more natural approach is to let

$$\tau(\alpha) = f[\alpha]\cdot\lambda(\alpha) = \{a\cdot\lambda(\alpha) \; ; \; a \in f[\alpha]\}$$

for $\alpha \in H$, so that $\tau : H \to \text{sb } X$, and consider $\int_A d\tau$. This involves working in $X' = \text{sb } X$, rather than X itself, with addition and topology on X' induced by the corresponding notions on X (see example 3 in Chapter I, sec.0.1). Our definition of $\int_A f\cdot d\lambda$ in 1.2(3) above then corresponds to accepting $\int_A d\tau$ only when it consists of a single point. This actually requires checking, but the work is straightforward. We should observe that, even when X is a group, X' is only a semigroup, a fact which offers another motivation for studying semigroup-valued measures.

In order to get the analogue of the classical theorem, in the case of the reals, that the integral of a bounded measurable function whith respect to a finite measure exists, we now turn our attention to the concept of measurable function.

2. Measurable and Partitionable Functions.

In the concept of a measurable function, the measure itself plays only a secondary role. The significant roles are played by the family of measurable sets and the family of null sets. With this in mind, then, we suppose

H is a σ-field of subsets of S and

H_o is a subfamily of H such that

(i) $\alpha \subset \beta \in H_o \Rightarrow \alpha \in H_o$ and

(ii) $\alpha_n \in H_o$ for $n \in \omega \Rightarrow \bigcup_{n\in\omega} \alpha_n \in H_o$.

2.1. Definitions.

(1) g is a simple function on S iff range g is finite and, for every $x \in$ range g , $g^{-1}[\{x\}] \in H$.

(2) g is a step function on S iff range g is countable and, for every $x \in$ range g , $g^{-1}[\{x\}] \in H$.

There are three main extensively used notions of measurable function.

2.2. Definitions. For any $f : S \to X_1$,

(1) When X_1 is the real line,

f is measurable iff, for every closed $\alpha \subset X$, $f^{-1}[\alpha] \in H$.

(2) When X_1 is a Banach space,

f is Bochner measurable iff there exist $S_o \in H_o$ and simple functions $g_n : S \to X_1$, for $n \in \omega$, such that

$$f(s) = \lim_n g_n(s) \qquad \text{for } s \in S \setminus S_o .$$

(3) When X_1 is a locally convex vector space,

f is weakly measurable iff, for every continuous linear functional ψ on X_1 , $\psi \circ f$ is measurable.

One readily verifies that the above three definitions agree when X_1 is the real line, but they are different in general. The next definition, besides extending the concept of measurability to multiple-valued functions and more general range spaces X_1 , brings the above approaches under a unified point of view. To avoid confusion with the existing literature, we use the term 'partitionable' instead of 'measurable'.

2.2. Definition.

For any uniform space X_1 and multiple-valued function f on S to X_1,

f is partitionable iff, for every element U of the uniformity for X_1, there exists a countable, disjoint $P \subset H$ such that

(i) $S \setminus \bigcup_{\alpha \in P} \alpha \in H_0$ and

(ii) for every $\alpha \in P$, $f[\alpha] \times f[\alpha] \subset U$.

2.3. Definition.

When λ is an outer measure on S, the notions of λ-measurable and λ-partitionable function are obtained from the above definitions by letting $H = \mathcal{M}_\lambda$ and $H_0 = \mathcal{N}_\lambda$.

In order to discuss the main properties of measurable and partitionable functions, we need the following concepts.

2.4. Definitions.

For any multiple-valued function f on S to a uniform space X_1,

(1) f is almost single-valued iff there exists $S_0 \in H$ such that

$f / (S \setminus S_o)$ is single-valued.

(2) f is <u>almost separably-valued</u> iff there exists $S_o \in H_o$ such that $f[S \setminus S_o]$ is separable.

(3) f is <u>quasi bounded</u> iff, for every element U of the uniformity for X_1, there exist $S_o \in H_o$ and a countable $E \subset X_1$ such that $f[S \setminus S_o] \subset U[E]$.

As simple consequences of the definitions, we have the following results.

2.5. <u>Lemma</u>. Let X_1 be a uniform space with uniformity $\mathcal{U}$ and f be a partitionable, multiple-valued function on S to X_1. Then

(1) for any $E \subset X$ and $U \in \mathcal{U}$, there exists $\beta \in H$ with

$$f^{-1}[E] \subset \beta \subset f^{-1}[U[E]] .$$

(2) if $U_n \in \mathcal{U}$ for $n \in \omega$ and $E = \bigcap_{n\in\omega} U_n[E]$ then

$$f^{-1}[E] \in H .$$

<u>Proof</u>:

(1) Choose a countable $P \subset H$, such that

$$S_o = S \setminus \bigcup_{\alpha\in P} \alpha \in H_o \qquad \text{and}$$

$$f[\alpha] \times f[\alpha] \subset U \qquad \text{for} \quad \alpha \in P .$$

Let

$$F = \{\alpha \in P : f[\alpha] \cap E \neq \emptyset\} \qquad \text{and}$$

$$\beta = \bigcup_{\alpha \in F} \alpha \cup (S_o \cap f^{-1}[E]) .$$

(2) is an immediate consequence of (1) . I

2.6. Theorem. Let X_1 be a uniform space with uniformity $\mathcal{U}$ and f be a partitionable, multiple-valued function on S to X_1 . Then

(1) f is quasi-bounded;

(2) for every $U \in \mathcal{U}$, there exist $S_o \in H_o$ and a step function $g : S \to X_1$ such that, for every $s \in S \setminus S_o$,

$$f(s) \subset U[g(s)] \quad ;$$

(3) when there is a countable base for $\mathcal{U}$ (i.e., X_1 is metrizable), there exists $S_o \in H_o$ such that $f / (S \setminus S_o)$ is the uniform limit of a sequence of step functions.

(4) When X_1 is a topological vector space, for every translate E of a closed, convex neighborhood of $\underline{0}$ in X_1 , $f^{-1}[E] \in H$.

Proof: (1) and (2) are straight forward consequences of the definitions; (3) follows from (2); (4) follows from lemma 2.5(2) after noting that, for any closed, convex, neighborhood U of $\underline{0}$ in X_1, if $E = x + U$ then

$$E = \bigcap_{n\varepsilon\omega_+} (E + \tfrac{1}{n}U) \quad ,$$

for, if $y = x + u_n + \frac{1}{n}v_n$ with $u_n, v_n \; \varepsilon \; U$ for $n \; \varepsilon \; \omega_+$, then

$$\frac{y - x}{1 + \frac{1}{n}} = \frac{1}{1 + \frac{1}{n}} u_n + \frac{(\frac{1}{n})}{1 + \frac{1}{n}} v_n \quad \varepsilon \quad U$$

and, letting $n \to \infty$, we get $y - x \; \varepsilon \; U$ so $y \; \varepsilon \; x + U = E$. ∎

The relations between partitionability and the various classical notions of measurability are summarized in the following theorems.

2.7. Theorem. Let X_1 be a metric space and f be a multiple-valued function on S to X_1 . Then f is partitionable iff

(i) f is almost single-valued,

(ii) f is almost separably valued, and

(iii) for every closed $E \subset X_1$, $f^{-1}[E] \; \varepsilon \; H$.

<u>Proof</u>: This is a straight forward consequence of the definitions and lemma 2.5(2) since there is a countable base for the uniformity for X_1 and every closed set E satisfies the hypothesis of 2.5(2). ∎

2.8. <u>Theorem</u>. Let $f : S \to X_1$

(1) When X_1 is a Banach space,

f is partitionable iff f is Bochner measurable.

(2) When X_1 is a locally convex vector space endowed with the weak topology,

f is partitionable iff f is weakly measurable.

<u>Proof</u>:

(1) Let X_1 be a Banach space.

(a) Suppose f is partitionable and, for $n \in \omega$, let $Z_n \in H_o$ and P_n be a countable, disjoint subfamily of H which covers $S \setminus Z_n$ and, for every $\alpha \in P_n$, $\operatorname{diam} f[\alpha] < \frac{1}{n+1}$. Let

$$S_o = \bigcup_{n\in\omega} Z_n \quad ,$$

$$P_n = \{\alpha(n, k) \; ; \; k \in \omega\} \quad ,$$

$$T_n = \{0, \cdots, n\}^n \quad .$$

For any $j \in T_n$, let

$$\beta(j) = \bigcap_{i=0}^{n} \alpha(i, j_i)$$

and, for $m = 0, \cdots, n-1$,

$$\beta(j / m) = \bigcap_{i=0}^{m} \alpha(i, j_i) \setminus \bigcap_{i=0}^{m+1} \alpha(i, j_i) .$$

Then, $\{\beta(j / m) ; j \in T_n , m = 0, \cdots, n\}$ is a finite, disjoint subfamily of H. If $\beta(j / m) \neq \emptyset$, let $x(j / m) \in f[\beta(j / m)]$ and set, for $s \in S$,

$$g_n(s) = \begin{cases} x(j / m) & \text{if } s \in \beta(j / m) \text{ for some } m \leq n \\ y & \text{otherwise} \end{cases}$$

where $y \in X_1$ is arbitrary. Then g_n is a simple function and, for $s \in S \setminus S_o$, we must have $f(s) = \lim_n g_n(s)$. To check this, given $\varepsilon > 0$, choose $N \in \omega$ with $\frac{1}{N} < \varepsilon$ and let k be the sequence in ω with $s \in \alpha(i, k_i)$ for every $i \in \omega$. For any $n \in \omega$ with $n > \max\{N, k_o, \cdots, k_N\}$, we must have $s \in \beta(j / m)$ for some $j \in T_n$ and $m \in \omega$ with $N \leq m \leq n$ so $\beta(j / m) \subset \alpha(N, j_N)$,

$$\operatorname{diam} f[\beta(j / m)] < \frac{1}{N} < \varepsilon \quad , \text{ and } \quad \|f(s) - g_n(s)\| < \varepsilon \quad .$$

Thus, f is Bochner measurable.

(b) Suppose f is Bochner measurable. Then f satisfies (i), (ii), (iii) of theorem 2.7 so f is partitionable.

(2) Let X_1 be a locally convex vector space endowed with the weak topology.

(a) Suppose f is partitionable. Then, for any continuous linear functional ψ on X_1 and any closed interval α,

$$\psi^{-1}[\alpha] = \bigcap_{n\varepsilon\omega} (\psi^{-1}[\alpha] + U_n)$$

where $U_n = \psi^{-1}[(-\frac{1}{n}; \frac{1}{n})]$ is a neighborhood of $\underline{0}$ in X_1. Hence, by lemma 2.5(2),

$(\psi \circ f)^{-1}[\alpha] = f^{-1}[\psi^{-1}[\alpha]] \varepsilon H$. Thus f is weakly measurable.

(b) Suppose f is weakly measurable. For any continuous linear functional ψ on X_1, let $U = \psi^{-1}[(-1; 1)]$. Then a countable family of translates of U, say $\{U_n; n \varepsilon \omega\}$, covers X_1. Letting,

$$\alpha_o = f^{-1}[U_n] \ , \quad \alpha_{n+1} = f^{-1}[U_{n+1}] \setminus \bigcup_{i=0}^{n} \alpha_i \ ,$$

we get $\alpha_n \varepsilon H$ and $f[\alpha_n] \varepsilon U_n$, for $n \varepsilon \omega$, and $\{\alpha_n ; n \varepsilon \omega\}$ is a countable, disjoint cover of S. Since such U form a subbase for the neighborhoods of $\underline{0}$ in X_1 and hence $\bigcup_{n\varepsilon\omega} (U_n \times U_n)$ a subbase for the uniformity for X_1, we conclude that f is partitionable. ∎

3. Properties of Partitionable Functions.

Again in this section we suppose

H is a σ-field of subsets of S and

H_o is a subfamily of H such that

(i) $\alpha \subset \beta \in H_o \Rightarrow \alpha \in H_o$ and

(ii) $\alpha_n \in H_o$ for $n \in \omega \Rightarrow \bigcup_{n\in\omega} \alpha_n \in H_o$.

3.1. Definitions.

Given a uniform space X_1 with uniformity $\mathcal{U}$,

(1) for any $U \in \mathcal{U}$,

$$U' = \{(D, E) : D \subset X_1 , E \subset X_1 , D \subset U[E] \text{ and } E \subset U[D]\}$$

(2) The uniformity on sb X_1 induced by $\mathcal{U}$ is the uniformity generated by $\{U' ; U \in \mathcal{U}\}$.

3.2. Theorem.

Let X_1 be a uniform space and, for every $n \in \omega$, let f_n be a partitionable, multiple-valued function on S to X_1 . If, for every $s \in S$, $g(s) \in$ sb X_1 and $g(s) = \lim_n f_n(s)$ in the induced uniform topology on sb X_1 , then g is a partitionable function.

<u>Proof:</u> Given an element U of the uniformity for X_1, choose closed elements V_1, V with $V_1 \circ V_1 \circ V_1 \subset U$ and $V \circ V \circ V \subset V_1$. For each $n \in \omega$, choose $Z_n \in H_o$ and countable, disjoint $P_n \subset H$ such that P_n covers $(S \setminus Z_n)$ and $f_n[\alpha] \times f_n[\alpha] \subset V$ for every $\alpha \in P_n$. Let

$$S_o = \bigcup_{n\in\omega} Z_n \qquad \text{and}$$

$$P' = \bigcup_{n\in\omega} P_n \quad .$$

Then $S_o \in H_o$ and P' is countable. Given $N \in \omega$ and $\beta \in P_N$, let

$$Q_n(\beta) = \{\alpha \in P_n : f_n[\alpha] \subset V_1[f_N[\beta]]\}$$

$$\beta' = \bigcap_{n=N}^{\infty} \bigcup_{\alpha\in Q_n(\beta)} \alpha \quad .$$

Then $\beta' \in H$ and we note that

(i) $g[\beta'] \times g[\beta'] \subset U$.

Indeed, if $s \in \beta'$ and $x \in f_N[\beta]$, we have for every $n \in \omega$ with $n \geq N$,

$f_n(s) \subset V_1[f_N[\beta]] \subset V_1 \circ V[x]$ hence

$g(s) \subset V_1 \circ V[x]$ and

$g[\beta'] \times g[\beta'] \subset V_1 \circ V_1 \circ V \circ V \subset U$.

(ii) $S \setminus S_o \subset \bigcup_{\beta\in P'} \beta'$.

Indeed, if $s \in S \setminus S_o$ then, for every $n \in \omega$, there exists $\alpha_n \in P_n$ with $s \in \alpha_n$. Moreover, there exists $N \in \omega$ such that for every $n \geq N$, we have

$$f_n(s) \subset V[g(s)] \quad \text{and} \quad g(s) \subset V[f_n(s)] \ .$$

Let $\beta = \alpha_N$. Then, for $n \geq N$,

$$f_n[\alpha_n] \subset V[f_n(s)] \subset V \circ V[g(s)]$$

$$\subset V \circ V \circ V[f_N(s)] \subset V_1[f_N[\beta]]$$

so $\alpha_n \in Q_n(\beta)$ and hence $s \in \beta'$.

By taking portions of the β' , we construct a countable, disjoint $P \subset H$ which refines $\{\beta' \ ; \ \beta \in P'\}$ so that, for every $\alpha \in P$, $g[\alpha] \times g[\alpha] \subset U$. I

3.3. Theorem. Let X_1 be a uniform semigroup. If f and g are partitionable, multiple-valued functions on S to X_1 then $f + g$ is partitionable.

<u>Proof:</u> Let $h = f + g$. Given an element U of the uniformity for X_1 , choose an element V so that

$$(x_1, y_1) \in V \text{ and } (x_2, y_2) \in V \Rightarrow (x_1+x_2, y_1+y_2) \in U$$

and a countable, disjoint $P \subset H$ such that, for every $\alpha \in P$,

$$f[\alpha] \times f[\alpha] \subset V \quad \text{and} \quad g[\alpha] \times g[\alpha] \subset V$$

and $(S \setminus \bigcup_{\alpha \in P} \alpha) \in H_o$. Then, for $x_1, y_1 \in f[\alpha]$ and $x_2, y_2 \in g[\alpha]$, we have $(x_1, y_1) \in V$ and $(x_2, y_2) \in V$ hence $(x_1+x_2, y_1+y_2) \in U$. Thus, $(f[\alpha] + g[\alpha]) \times (f[\alpha] \times g[\alpha]) \subset U$. I

We conclude this section with a characterization, when X_1 is a topological vector space, of partitionable functions, in terms of weakly measurable ones. This generalizes a well known theorem of Pettis about when a weakly measurable function with values in a Banach space is Bochner measurable. For this, we first consider a few topological concepts.

3.4. <u>Definitions</u>.

Let X_1 be a topological vector space and X' be its topological dual.

(1) for $E \subset X_1$,

$$E^\circ = \{\psi \in X' : |\psi(x)| \leq 1 \text{ for every } x \in E\} .$$

(2) for $L \subset X'$,

$$L^\circ = \{x \in X_1 : |\psi(x)| \leq 1 \text{ for every } \psi \in L\} .$$

(3) for any closed, convex, symmetric neighborhood U of $\underline{0}$ in X_1 and any $x \in X_1$,

$$\|x\|_U = \inf\{r \geq 0 : x \in r\,U\} .$$

Note: (i) $\|\cdot\|_U$ is a semi-norm on X_1 ;

(ii) $U = \{x \in X_1 : \|x\|_U \leq 1\} = U^{\circ\circ}$ (Hahn-Banach) ;

(iii) for any subspace E of X_1 which is separable in $\|\cdot\|_U$, there is a countable $L \subset U^\circ$ such that

$$U \cap E = L^\circ \cap E .$$

To see (iii), let $\{x_n ; n \in \omega\}$ be dense in E in $\|\cdot\|_U$ and, by the Hahn-Banach theorem, take $\psi_n \in U^\circ$ so that $\psi_n(x_n) = \|x_n\|_U$ and let $L = \{\psi_n ; n \in \omega\}$.

3.5. Lemma. Let X_1 be a topological vector space and f be a quasi bounded, multiple-valued function on S to X_1. Then, for every closed, convex, symmetric neighborhood U of $\underline{0}$ in X_1, there exists $S_o \varepsilon H_o$ and, for any $x \varepsilon X_1$, a countable $L \subset U$ with

$$(S \setminus S_o) \cap f^{-1}[x + U] = (S \setminus S_o) \cap \bigcap_{\psi \varepsilon L} f^{-1}[x + \{\psi\}^\circ] .$$

Proof: Since, f is quasi bounded, for every $n \varepsilon \omega_+$, there exists $Z_n \varepsilon H_o$ such that a countable number of translates of $\frac{1}{n} U$ cover $f[S \setminus Z_n]$. Let

$$S_o = \bigcup_{n \varepsilon \omega_+} Z_n .$$

Then $S_o \varepsilon H_o$ and $f[S \setminus S_o]$ is separable in $\|\cdot\|_U$. For any $x \varepsilon X_1$, let E be the subspace spanned by $f[S \setminus S_o] - x$. Then E is also separable in $\|\cdot\|_U$, hence by remark (iii) above, there exists a countable $L \subset U^\circ$ with

$$U \cap (f[S \setminus S_o] - x) = L^\circ \cap (f[S \setminus S_o] - x)$$

or

$$f[S \setminus S_o] \cap (x + U) = f[S \setminus S_o] \cap (x + L^\circ) .$$

Since

$$x + L^\circ = x + \bigcap_{\psi \varepsilon L} \{\psi\}^\circ = \bigcap_{\psi \varepsilon L} (x + \{\psi\}^\circ) ,$$

we conclude

$$(S \setminus S_o) \cap f^{-1}[x + U] = (S \setminus S_o) \cap \bigcap_{\psi \varepsilon L} f^{-1}[x + \{\psi\}^\circ] . \text{ I}$$

3.6. Theorem. Let X_1 be a locally convex vector space and $f : S \to X_1$. Then f is partitionable iff f is quasi bounded and weakly measurable.

Proof: Clearly, from theorem 2.6, if f is partitionable then f is quasi bounded and weakly measurable. Now, suppose f is quasi-bounded and weakly measurable. Then, for any closed, convex, symmetric neighborhood U of $\underline{0}$, there exist $x_n \varepsilon X_1$ and $S_o \varepsilon H_o$ such that

$$f[S \setminus S_o] \subset \bigcup_{n \varepsilon \omega} (x_n + U) .$$

Since, for each $n \varepsilon \omega$ and continuous linear functional ψ on X_1, $f^{-1}[x_n + \{\psi\}^\circ] \varepsilon H$, we have, by lemma 3.5, that $\alpha_n = f^{-1}[x_n + U] \varepsilon H$. Letting $\beta_o = \alpha_o$, $\beta_{n+1} = \alpha_{n+1} \setminus \bigcup_{i=0}^{n} \alpha_i$, $P = \{\beta_n ; n \varepsilon \omega\}$, we see that P is countable, disjoint, covers $(S \setminus S_o)$ since f is single-valued, and $f[\beta] - f[\beta] \subset U$ for every $\beta \varepsilon P$.

Thus, f is partitionable. I

As a corollary of the above and the results in section 2, we have the following.

3.7. Theorem. Let X_1 be a locally convex vector space and $f : S \to X_1$. Then f is partitionable iff f is quasi bounded and any one of the following holds:

(1) f is weakly measurable;

(2) for every half-plane $E \subset X_1$, $f^{-1}[E] \in H$;

(3) for every closed, convex neighborhood E of a point in X_1, $f^{-1}[E] \in H$;

(4) there exists a base $\mathcal{B}$ for the neighborhoods of $\underline{0}$ in X_1 such that, for every $U \in \mathcal{B}$ and $x \in X_1$, $f^{-1}[x + U] \in H$;

(5) for every neighborhood U of $\underline{0}$ in X_1 there exists $S_o \in H_o$ and a step function $g : S \to X_1$ such that $g(s) - f(s) \in U$ for every $s \in S \setminus S_o$.

4. <u>Existence of the Integral</u>.

We shall now consider conditions under which $\int f\cdot d\lambda$ exists. We suppose

S is an abstract space;

H is a ring of subsets of S ;

H_o is a subfamily of H such that

$$\alpha \subset \beta \in H_o \quad \Rightarrow \quad \alpha \in H_o \ ;$$

X_1 is a uniform space with uniformity $\mathcal{U}_1$;

X_2 is a commutative, topological semigroup;

X is a commutative, Hausdorff, uniform semigroup with uniformity $\mathcal{U}$;

$\cdot$: $(a, b) \in X_1 \times X_2 \rightarrow a\cdot b \in X$ with

$$a\cdot(b_1 + b_2) = a\cdot b_1 + a\cdot b_2 \quad \text{and} \quad a\cdot\underline{0} = \underline{0} \ ;$$

f is a multiple-valued function on S to X_1 ;

λ : $H \rightarrow X_2$ with $\lambda(\alpha) = \underline{0}$ for every $\alpha \in H_o$.

4.1. Definitions.

(1) f behaves almost as a bounded function iff, for every neighborhood V of $\underline{0}$ in X, there exist $S_o \in H_o$ and a neighborhood V_2 of $\underline{0}$ in X_2 such that if $a_i \in f[S \setminus S_o]$ and $b_i \in V_2$ for $i = 0, \cdots, n$ and, for every $A \subset \{0, \cdots, n\}$,

$$\sum_{i \in A} b_i \in V_2$$

then

$$\sum_{i=0}^{n} a_i b_i \in V .$$

(2) λ behaves as a bounded finitely additive measure iff

(i) λ is finitely additive on H ;

(ii) λ is s-bounded (see definition 5.1 in Chap. I) ;

(iii) for every $U \in \mathcal{U}$ there exists $U_1 \in \mathcal{U}_1$ such that if $a_i, b_i \in X_1$ with $(a_i, b_i) \in U_1$ and $\alpha_i \in H$ with $\alpha_i \cap \alpha_j = \emptyset$ for $i,j = 0, \cdots, n$ and $i \neq j$ then

$$\left(\sum_{i=0}^{n} a_i \cdot \lambda(\alpha_i), \sum_{i=0}^{n} b_i \cdot \lambda(\alpha_i)\right) \in U .$$

Our fundamental existence theorem is then the following.

4.2. Theorem. Suppose f is partitionable and behaves almost as a bounded function, λ behaves as a bounded finitely additive measure, and X is complete. Then

(i) $\int_A f\cdot d\lambda \in X$ for every $A \subset S$;

(ii) if $\mu(A) = \int_A fd\mu$ then μ is an H_σ-outer measure.

Proof:

(i) We shall show that the sums occuring in the definition of the integral form a Cauchy net in X . Given $U \in \mathcal{U}$, since X is a uniform semigroup, there exists a neighborhood V of $\underline{0}$ in X such that, for every $x \in X$ and $v \in V$, $(x, x+v) \in U$. Let V_2 be a neighborhood of $\underline{0}$ in X_2 having the property stated in definition 4.1(1), let $U_1 \in \mathcal{U}_1$ be as in definition 4.1(2), and $A \subset S$.

(1) There exists $(P, \Delta) \in \mathfrak{P}(A)$ such that, for every $(P', \Delta') \in \mathfrak{P}(A)$ with $(P, \Delta) << (P', \Delta')$, we have

(a) $\alpha \in H$, F finite, $F \subset P \setminus \Delta(P)$,

$$\alpha \subset \bigcup_{\beta \in F} \beta \quad \Rightarrow \quad \lambda(\alpha) \in V_2$$

(b) $\alpha \in H, \ \alpha \subset \bigcup_{\beta\in\Delta(P)} \beta \setminus \bigcup_{\beta\in\Delta'(P')} \beta \quad \Rightarrow \quad \lambda(\alpha) \in V_2$

To see this, for $n \in \omega$, choose neighborhoods W_n of $\underline{0}$ in X_2 so that

$$W_o + W_o + W_o \subset V_2 \quad \text{and} \quad W_{n+1} + W_{n+1} \subset W_n .$$

Since λ is s-bounded, for any $P \in \mathcal{P}(A)$, choose a finite $\Delta_n(P) \subset P$ so that $\Delta_n(P) \subset \Delta_{n+1}(P)$ and $\alpha \in H$, F finite, $F \subset P \setminus \Delta_n(P)$, $\alpha \subset \bigcup_{\beta\in F} \beta \Rightarrow \lambda(\alpha) \in W_n$. If statement (1) were not true, there would exist $P_n \in \mathcal{P}(A)$ with $P_n < P_{n+1}$ and $\alpha_n \in H$, for $n \in \omega$, such that

$$\alpha_n \subset B_n \setminus B_{n+1} \quad \text{and} \quad \lambda(\alpha_n) \notin V_2$$

where $B_n = \bigcup_{\beta\in\Delta_n(P_n)} \beta$.

Let $\gamma_n = \bigcap_{i=0}^{n} B_i$.

Then

$$\alpha_n = (\alpha_n \cap \gamma_n \setminus \gamma_{n+1}) \cup (\alpha_n \setminus \gamma_n) .$$

The $(\alpha_n \cap \gamma_n \setminus \gamma_{n+1})$ form a disjoint sequence in H hence, for large n, $\lambda(\alpha_n \cap \gamma_n \setminus \gamma_{n+1}) \in W_o$. On the other hand,

$$\alpha_n \setminus \gamma_n = (\alpha_n \setminus B_o) \cup (\alpha_n \cap B_o \setminus B_1)$$

$$\cup \cdots \cup (\alpha_n \cap \gamma_{n-2} \setminus B_{n-1})$$

so

$$\lambda(\alpha_n \setminus \gamma_n) \in W_o + W_1 + \cdots + W_{n-1} \subset W_o + W_o \ .$$

Thus, for large n , $\lambda(\alpha_n) \in W_o + W_o + W_o \subset V_2$ contradicting the choice of α_n above.

Now, since f is partitionable, choose $(P, \Delta) \in \mathfrak{P}(A)$ satisfying (a) and (b) in statement (1) above and such that, for every $\alpha \in P$, we have $f[\alpha] \times f[\alpha] \subset U_1$ or $\alpha \in H_o$. We shall now check that, for any $(P', \Delta') \in \mathfrak{P}(A)$ with $(P, \Delta) << (P', \Delta')$ and any choice functions ξ and η for f ,

$$\Big(\sum_{\alpha\in\Delta(P)} \xi_\alpha \cdot \lambda(\alpha), \sum_{\beta\in\Delta'(P')} \eta_\beta \cdot \lambda(\beta)\Big) \in U \circ U \circ U \ .$$

To this end, let

$$B = \bigcup_{\alpha\in\Delta(P)} \alpha \cap \bigcup_{\beta\in\Delta'(P')} \beta$$

$$Q_\alpha = \{\beta \in \Delta'(P') : \beta \subset \alpha \text{ and } \beta \not\subset H_o\} \ .$$

Then

$$\sum_{\alpha\in\Delta(P)} \xi_\alpha \cdot \lambda(\alpha) = \sum_{\alpha\in\Delta(P)} \xi_\alpha \cdot \lambda(\alpha \cap B) + \sum_{\alpha\in\Delta(P)} \xi_\alpha \cdot \lambda(\alpha \setminus B)$$

$$= \sum_{\alpha\in\Delta(P)} \sum_{\beta\in Q_\alpha} \xi_\alpha \cdot \lambda(\beta) + \sum_{\alpha\in\Delta(P)} \xi_\alpha \cdot \lambda(\alpha \setminus B)$$

and

$$\sum_{\beta\in\Delta'(P')} \eta_\beta\cdot\lambda(\beta) = \sum_{\alpha\in\Delta(P)} \sum_{\beta\in Q_\alpha} \eta_\beta\cdot\lambda(\beta) + \sum_{\beta\in\Delta'(P')} \eta_\beta\cdot\lambda(\beta \setminus B)$$

Now, for any finite $F \subset \Delta(P)$, by 1(b) ,

$$\sum_{\alpha\in F} \lambda(\alpha \setminus B) \in V_2$$

hence

$$\sum_{\alpha\in\Delta(P)} \xi_\alpha\cdot\lambda(\alpha \setminus B) \in V \quad .$$

Similarly, because of 1(a),

$$\sum_{\beta\in\Delta'(P')} \eta_\beta\cdot\lambda(\beta \setminus B) \in V \quad .$$

On the other hand, for $\alpha \in \Delta(P)$ and $\beta \in Q_\alpha$, we have $(\xi_\alpha, \eta_\beta) \in U_1$. Hence

$$\Big(\sum_{\alpha\in\Delta(P)} \sum_{\beta\in Q_\alpha} \xi_\alpha\cdot\lambda(\beta), \sum_{\alpha\in\Delta(P)} \sum_{\beta\in Q_\alpha} \eta_\beta\cdot\lambda(\beta)\Big) \in U \quad .$$

We conclude therefore,

$$\Big(\sum_{\alpha\in\Delta(P)} \xi_\alpha\cdot\lambda(\alpha), \sum_{\beta\in\Delta'(P')} \eta_\beta\cdot\lambda(\beta)\Big) \in U \circ U \circ U \ .$$

(ii) is an immediate consequence of (i) and theorem 4.3 in Chapter I. ▌

As a corollary we have the following.

4.3. Theorem. Let X be a locally convex vector space, λ be a non-negative, finitely additive function on H, f be a multiple-valued function on S to X, and

$$C = \text{closed convex hull of } f[S] .$$

If $\lambda(S) < \infty$, f is partitionable, and C is bounded and complete then

$$\int_S f \cdot d\lambda \in \lambda(S) \cdot C .$$

Proof: Let $X_1 = X$, $X_2 = [0, \infty)$ and '$\cdot$' be ordinary multiplication of a vector in X by a scalar. Then apply theorem 4.2 after noting that

(i) C is bounded $\Rightarrow$ f behaves as a bounded function

(ii) $\lambda(S) < \infty$ $\Rightarrow$ λ behaves as a bounded, finitely additive measure.

To check (i), given any neighborhood U of $\underline{0}$ in X, choose $\varepsilon > 0$ so that $r \cdot C \subset U$ whenever $|r| < \varepsilon$. Then, for $x_i \in f[S]$ and $t_i \in X_2$ with $0 < r = \sum_{i=0}^{n} t_i < \varepsilon$, we have

$$\sum_{i=0}^{n} t_i \cdot x_i = r \cdot \sum_{i=0}^{n} \frac{t_i}{r} \cdot x_i \in r \cdot C \subset U .$$

To check (ii), given any convex neighborhood U of $\underline{0}$ in X, if $\lambda(S) \neq 0$, let $U_1 = \frac{1}{\lambda(S)} \cdot U$. Then for any $x_i, y_i \in X$ with $x_i - y_i \in U_1$ and disjoint $\alpha_i \in H$ for $i = 0, \dots, n$ with $A = \bigcup_{i=0}^{n} \alpha_i$, we have

$$\sum_{i=0}^{n} x_i \, \lambda(\alpha_i) - \sum_{i=0}^{n} y_i \, \lambda(\alpha_i)$$

$$= \lambda(A) \cdot \sum_{i=0}^{n} \frac{\lambda(\alpha_i)}{\lambda(A)} (x_i - y_i) \in \lambda(A) \cdot U_1 \subset U . \quad \blacksquare$$

4.4. Remarks.

(1) Theorem 4.2 above solves simultaneously the problem of producing an outer measure and that of producing a Lebesgue type integral from a given finitely additive measure. Note that theorem 5.2 in Chapter I is virtually a special case of theorem 4.2 above, obtained by letting $X_1 = \{1\}$, $X_2 = X$ and $1 \cdot b = b$ for $b \in X$.

(2) The proof of statement (1) in theorem 4.2 above is essentially that of theorem 5.2 in Chapter I. When λ is σ-additive, statement (1) is trivial.

(3) When X_2 is a complete uniform semigroup, by theorem 6.1 in Chapter I, letting $\lambda^*(A) = \int_A d\lambda$, we get an outer measure λ^*.

Then under the conditions of theorem 4.2 above, one can check that

$$\int_A f\cdot d\lambda^* = \int_A f\cdot d\lambda \quad .$$

Thus, theorem 4.2 enables us to bypass the construction of λ^* in the production of the integral. Only when λ is σ-additive, is λ^* an extension of λ . Thus, a theorem like 4.2 cannot be obtained through a Daniell approach to integration.

5. Integrable Functions.

We shall now prove our version of the Lebesgue dominated convergence theorem. Our approach is through the notion of uniform integrability and yields, even on the real line, a result somewhat stronger than the classical one. Our basic set up is the same as in section 4.

5.1. Definitions.

(1) F is a uniformly s-integrable family iff F is a family of partitionable, multiple-valued functions on S to X_1 and, for every $U \in \mathcal{U}$, there exist $U_1 \in \mathcal{U}_1$, a neighborhood V_2 of $\underline{0}$ in X_2 and $A_f \in H$ for $f \in F$ such that

(i) $\left(\int_B f\cdot d\lambda, \int_{B\cap A_f} f\cdot d\lambda\right) \in U$ for every $B \subset S$;

(ii) if $a_i \in f[A_f]$, $b_i \in V_2$ for $i = 0, \cdots, n$ and, for every $J \subset \{0, \cdots, n\}$, $\sum_{i\in J} b_i \in V_2$ then

$$\sum_{i=0}^{n} a_i\cdot b_i \in U[\underline{0}] ;$$

(iii) if $a_i, b_i \in X_1$ with $(a_i, b_i) \in U_1$ and $\alpha_i \in H$ with $\alpha_i \cap \alpha_j = \emptyset$ for $i,j = 0, \cdots, n$ and $i \neq j$

then

$$\left(\sum_{i=0}^{n} a_i \cdot \lambda(\alpha_i \cap A_f),\ \sum_{i=0}^{n} b_i \cdot \lambda(\alpha_i \cap A_f)\right) \in U ;$$

(iv) if $\alpha_n \in H$, $\alpha_n \subset \bigcup_{f \in F} A_f$ and $\alpha_m \cap \alpha_n = \emptyset$ for $m, n \in \omega$ with $m \neq n$, then $\lim_n \lambda(\alpha_n) = \underline{0}$.

(2) f is s-<u>integrable</u> iff $\{f\}$ is a uniformly s-integrable family.

<u>Note</u>: If X_1 is a Banach space, $X_2 = [0; \infty]$, λ is a non-negative measure, and $\int \|f\| d\lambda < \infty$ then, for every $\varepsilon > 0$, there exists a $K \in \omega$ such that if $A = \{s : \frac{1}{K} < \|f(s)\| < K\}$ then $\int_{S \setminus A} \|f\| d\lambda < \varepsilon$ and $\lambda(A) < \infty$. Thus, a family of measurable functions dominated by such an f is a uniformly s-integrable family. In the above definitions, condition (ii) translates the idea that f is bounded on A and conditions (iii), (iv) the idea that $\lambda(A) < \infty$.

5.2. <u>Theorem</u>. Suppose H is a σ-field and $(H_o)_\sigma = H_o$, λ is finitely additive on H , $\{f_n; n \in \omega\} \cup \{g\}$ is a uniformly s-integrable family, $S_o \in H_o$ and, for every $s \in (S \setminus S_o)$, $g(s) = \lim_n f_n(s)$ in the induced uniform topology on sb X_1 . Then for every $B \subset S$,

$$\int_B g \cdot d\lambda = \lim_n \int_B f_n \cdot d\lambda .$$

Proof: Given a closed $V \in \mathcal{U}$, choose $U \in \mathcal{U}$ so that $U \circ U \circ U \subset V$ and $[(a, b) \in U$ and $(c, d) \in U \Rightarrow (a+c, b+d) \in V]$, and let $U_1, V_2, A_n = A_{f_n}, A_g$ be as in definition 5.1 above. Choose $V_1, W_1 \in \mathcal{U}_1$ so that $V_1 \circ V_1 \circ V_1 \subset U_1$ and $W_1 \circ W_1 \circ W_1 \subset V_1$. Since H is a σ-field and f_n, g are partitionable, there exist $\beta_n \in H_o$ and $\gamma_n \in H$ for $n \in \omega$ such that

$$\{s \in S \setminus \beta_n : f_n(s) \times g(s) \subset W_1\} \subset \gamma_n \subset \{s : f_n(s) \times g(s) \subset V_1\} .$$

Let

$$C_N = \bigcap_{n \geq N} \gamma_n \quad \text{and} \quad S_1 = S_o \cup \bigcup_{n \in \omega} \beta_n .$$

Then

$$S_1 \in H_o, \quad C_N \subset C_{N+1} \in H, \quad (S \setminus S_1) \subset \bigcup_{N \in \omega} C_N$$

and, by condition (iv) of 5.1, there exists $N \in \omega$ so that $\lambda(\alpha) \in V_2$ whenever

$$\alpha \in H, \quad \alpha \subset \bigcup_{n \in \omega} A_n \cup A_g, \quad \text{and} \quad \alpha \subset C_{N+k} \setminus C_N \quad \text{for } k \in \omega.$$

Therefore, since λ is finitely additive on H, by condition (ii) of 5.1 we have, for any $B \subset S$ and $n \in \omega$ with $n > N$,

$$\int_{B \cap A_n \setminus C_N} f_n \cdot d\lambda \ \varepsilon \ U[\underline{0}]$$

and $$\int_{B \cap A_g \setminus C_N} g \cdot d\lambda \ \varepsilon \ U[\underline{0}] \quad .$$

Let $D = C_N \cap (A_n \cup A_g)$. Then

$$\left(\int_B f_n \cdot d\lambda, \int_{B \cap D} f_n \cdot d\lambda\right) \ \varepsilon \ U \circ U \circ U \subset V$$

and $$\left(\int_B g \cdot d\lambda, \int_{B \cap D} g \cdot d\lambda\right) \ \varepsilon \ U \circ U \circ U \subset V \quad .$$

Now choose $P_o \ \varepsilon \ \mathcal{P}(B \cap D)$ so that, for every $\alpha \ \varepsilon \ P_o \setminus H_o$, we have $\alpha \subset D$, $f_n[\alpha] \times f_n[\alpha] \subset V_1$ and $g[\alpha] \times g[\alpha] \subset V_1$ and hence, since $\alpha \subset C_N$,

$$f_n[\alpha] \times g[\alpha] \subset V_1 \circ V_1 \circ V_1 \subset U_1 \quad .$$

Then, for any $P \ \varepsilon \ \mathcal{P}(B \cap D)$ with $P_o < P$, any finite $J \subset P \setminus H_o$, and any choice functions ξ for f_n and η for g, we have $(\xi_\alpha, \eta_\alpha) \ \varepsilon \ U_1$ for $\alpha \ \varepsilon \ J$ and hence, by condition (iii) of 5.1,

$$\left(\sum_{\alpha \varepsilon J} \xi_\alpha \cdot \lambda(\alpha \cap A_n), \sum_{\alpha \varepsilon J} \eta_\alpha \cdot \lambda(\alpha \cap A_n)\right) \ \varepsilon \ U$$

and $\left(\sum_{\alpha\varepsilon J} \xi_\alpha \cdot \lambda(\alpha \cap A_g \setminus A_n),\ \sum_{\alpha\varepsilon J} \eta_\alpha \cdot \lambda(\alpha \cap A_g \setminus A_n)\right) \varepsilon\ U$,

so $\left(\sum_{\alpha\varepsilon J} \xi_\alpha \cdot \lambda(\alpha),\ \sum \eta_\alpha \cdot \lambda(\alpha)\right) \quad \varepsilon \quad V$.

We conclude

$$\left(\int_{B\cap D} f_n \cdot d\lambda,\ \int_{B\cap D} g \cdot d\lambda\right) \varepsilon\ V$$

and therefore

$$\left(\int_B f_n \cdot d\lambda,\ \int_B g \cdot d\lambda\right) \varepsilon\ V \circ V \circ V\ . \qquad \mathbf{I}$$

5.3. <u>Remarks</u>.

(1) In the above theorem, we could replace the requirement that $\{f_n;\ n\ \varepsilon\ \omega\} \cup \{g\}$ be a uniformly s-integrable family by the weaker condition that g be s-integrable and the f_n be terminally uniformly s-integrable, i.e., for any $U\ \varepsilon\ \mathcal{U}$, conditions (i)-(iv) in 5.1 hold only for large n .

(2) To see that the above theorem is somewhat stronger than the Lebesgue dominated convergence theorem, let $S = [0;\ 1]$, λ be Lebesgue measure on S , $g(s) = 0$ for $s\ \varepsilon\ S$, and for $n\ \varepsilon\ \omega$,

$$f_n(s) = \begin{cases} n & \text{for} \quad s\ \varepsilon\ [\frac{1}{n+1};\ \frac{1}{n}] \\ 0 & \text{for} \quad s\ \varepsilon\ S \setminus [\frac{1}{n+1};\ \frac{1}{n}] \end{cases} .$$

Then the f_n are not dominated by any summable function whereas

$\{f_n ; n \in \omega\} \cup \{g\}$ is a uniformly s-integrable family.

5.4. Remark. We can introduce an $\mathcal{L}_1$-space in our present context even though we lack a vector structure or a norm. Let

$$\mathcal{L}_1 = \{f : f \text{ is s-integrable}\}$$

and, for any $U \in \mathcal{U}$, let

$$\hat{U} = \{(f, g) : f,g \in \mathcal{L}_1 \text{ and } (\int_B f \cdot d\lambda, \int_B g \cdot d\lambda) \in U \text{ for every } B \subset S\}$$

and $\hat{\mathcal{U}}$ be the uniformity on $\mathcal{L}_1$ generated by $\{\hat{U} ; U \in \mathcal{U}\}$.

6. Relation to Bochner and Pettis Integrals.

Within the general setting of section 4, let $X_1 = X$ be a topological vector space, $X_2 = [0; \infty]$, λ be a non-negative outer measure, $H = \mathcal{M}_\lambda$, $H_o = \mathcal{N}_\lambda$, and $f : S \to X_1$.

6.1. Definitions.

(1) When X is a Banach space,

f is Bochner integrable iff f is Bochner measurable and $\int \|f\| \, d\lambda < \infty$.

In this case, we set

$$(\text{Bochner})\text{-}\int f d\lambda = \lim_n \int g_n d\lambda$$

where g is any sequence of simple functions tending to f almost everywhere (this limit is independent of the choice of g_n) .

(2) When X is a locally convex space,

f is Pettis integrable iff f is weakly measurable and there exists $x \in X$ such that, for every continuous linear functional ψ on X ,

$$\psi(x) = \int (\psi \circ f) d\lambda .$$

In this case, we set

$$(\text{Pettis})\text{-}\int f d\lambda = x .$$

The relation of our integral to the above is given in the following theorems.

6.2. Theorems.

(1) When X is a Banach space,

if f is Bochner integrable then f is s-integrable and

$$\text{(Bochner)-}\int_B f d\lambda = \int_B f \cdot d\lambda \qquad \text{for any} \quad B \in \mathcal{M}_\lambda \ .$$

(2) When X is a locally convex space endowed with the weak topology,

(i) f is Pettis integrable iff $\int_S f \cdot d\lambda \in X$;

(ii) f is Pettis integrable over every $B \in \mathcal{M}_\lambda$ iff f is s-integrable;

(iii) $\text{(Pettis)-}\int_B f d\lambda = \int_B f \cdot d\lambda$ for any $B \in \mathcal{M}_\lambda$.

Proof: The identification of partitionability with the two notions of measurability was given in theorem 2.8 of this chapter. With the help of the remark after 5.1 and theorem 5.2, straight forward application of the definitions yields (1). We check (2) directly from the definitions and the fact that (2) holds when $X = (-\infty; \infty)$. I

6.3. Remarks.

(1) When X is a Banach space, f may be s-integrable even though f is not Bochner integrable. For example, let $S = \omega$, $X = \ell_\infty$, and for each $n \in \omega$, $\lambda(\{n\}) = 1$ and

$$f(n) = \frac{1}{n} \cdot 1_{\{n\}} .$$

Then $\|f(n)\| = \frac{1}{n}$ and hence $\int \|f\| d\lambda = \infty$. However, $\int f \cdot d\lambda = x \in X$ where $x_n = \frac{1}{n}$ for $n \in \omega$. Thus, the requirement that $\int \|f\| d\lambda < \infty$ is unnecessarily restrictive for the existence of the integral as the limit of sums in the norm topology.

(2) On the real line, unconditional and absolute convergence of sums coincide so the Lebesgue theory of integration may be considered as being involved with either or both of the above notions. In a Banach space, however, the two notions do not coincide in general. The Bochner theory of integration, with its emphasis on bounded variation, extends the Lebesgue theory along the lines of absolute convergence. Our theory on the other hand is primarily concerned with unconditional convergence. This permits us to extend the Bochner integral without abandoning the topology.

(3) When working with the weak topology on a locally convex space, theorem 6.2(2) shows that our theory coincides with that of Pettis. A well known extension of the Pettis integral is obtained by letting the integral x in definition 6.1(2) be a point in

X'^* , the algebraic dual of the topological dual of X . It is well known that X'^* is the completion of X in the weak topology. From our point of view, then, this amounts to working in the completion of X , instead of just X , when taking the limit of sums. This becomes all the more natural in the light of the completeness condition in the hypotheses of theorem 4.3 of this chapter.

(4) Our main existence theorem for the integral actually guarantees s-integrability. Even on the real line, there are hardly any general existence theorems which do not involve finiteness conditions.

REFERENCES

1. Bartle, R. G. , Dunford, N., Schwartz, J. Weak compactness and vector measures. Can. J. of Math., 7(1955), pp. 289-305.

2. Birkhoff, G. Integration of functions with values in a Banach space. Trans. A.M.S., 38(1935), pp. 357-378.

3. Bochner, S. Integration von Funktionen deren werte die Elemente eines Vectorraumes sind. Fund. Math., 20(1933), pp. 262-276.

4. Dunford, N. and Schwartz, J. Linear Operators, Vol. I, Chapter III, Interscience Publishers, Inc., New York, 1958.

5. Gelfand, I. M. Abstrakte Funktionen und Lineare Operatoren. Mat. Sbornik N.S. 4(1938), pp. 235-286.

6. Pettis, B. J. On integration in vector spaces. Trans. A.M.S. 44(1938), pp. 277-304.

7. Phillips, R. S. Integration in a convex linear topological space. Trans. A.M.S., 47(1940), pp. 114-145.

8. Price, G. B. The theory of integration. Trans. A.M.S., 47 (1940), pp. 1-50.

9. Rickart, C. E. Integration in a convex linear topological space. Trans. A.M.S., 52(1942), pp. 498-521.

10. Sion, M. Introduction to the methods of real analysis. Holt, Rinehart and Winston, New York, 1968.

11. _____________, Group-valued outer measures. Actes du Congres International des Mathematiciens, Nice 1970, vol. 2, pp. 589-593.

12. _____________, Lectures on vector-valued measures, University of British Columbia, 1969-70.

The idea of an integral of a vector-valued function is quite old as it occurs, for example, in the concept of a line integral.

The first explicit introduction of a Lebesgue type integral $\int f d\lambda$ for a vector-valued f is due to Bochner [3] who worked in a Banach space with the norm topology. The idea of using linear functionals to reduce the problem to the real-valued situation is due to Gelfand [5] and Pettis [6]. It was Gelfand who first considered the integral as a point in the double dual. Price [8] was first to study the problem when both f and λ are vector-valued.

The integral of a multiple-valued f was introduced by Birkhoff [2], using unconditional convergence of sums in a Banach space. This was generalized and refined by Phillips [7] and then by Rickart [9].

The idea of developing a Lebesgue type integral from a finitely additive λ, without first producing a measure, is found in Sion [10] for the real case and in Sion [11, 12] for the more general case. Partitionable functions were introduced by Sion [11, 12], but a closely related concept is also found in Rickart [9]. The concept of s-integrability is new.

CHAPTER III

DIFFERENTIATION

For most applications, the main interest in group or vector valued measures is motivated by integral representation theorems. These involve an equation of the form

$$U = \int f d\lambda$$

where U is a given element in some group, say an operator, and either f or λ or both are group valued. Invariably, the above equation is a special case of the more general one

$$\mu(A) = \int_A f d\lambda \qquad \text{for} \quad A \in H$$

where H is some family of subsets of a space S and μ is a group-valued measure with $U = \mu(S)$. Sometimes, as in the spectral theorem, the problem is to determine the measure λ, the function f being known from other considerations. More frequently, it is to find f once μ and λ are known. This is the problem of differentiation, f being the derivative of μ with respect to λ.

In this chapter, we shall develop a theory of differentiation which parallels the classical one of Lebesgue for real valued measures. Its key features are the notion of absolute continuity and the explicit definition of derivative as the limit of ratios. The main advantage of this approach is that it is conceptually elementary and very direct thereby permitting us to "see" where the derivative comes from. It enables us to incorporate the classical theory for real valued measures as a special case and frees us from reliance on any particular topology such as the weak topology on a Banach space. As a result, we develop a unified view of differentiation and bring under one roof a variety of seemingly unrelated integral representation theorems.

Our approach involves the use of a differentiation basis and, at first glance, this may appear to limit the scope of the theory. Indeed, this apparent limitation was a major reason for the loss of popularity of the Lebesgue view of differentiation in favor of the Radon-Nikodym view. We shall see however that the lifting theorem guarantees that a differentiation basis (in fact a Vitali basis) always exists. As a result, we can take advantage of the simplicity of the Lebesgue method without sacrificing any generality. As a matter of fact, we get a more powerful fundamental result by having a differentiation basis as a variable in the hypotheses. A great many theorems can be obtained from it by specifying the basis. This gives unity to the field and helps clarify the role of the lifting theorem in this area: it is needed when a basis cannot be obtained otherwise.

1. Absolute Continuity.

For the concepts of absolute continuity and singularity with respect to a base measure λ only the family of sets having λ-measure zero is involved. To emphasize this fact, in the following definitions, we replace the traditional base measure λ by a family H_o .

Throughout this section, we suppose

S is an abstract space;

H is a ring of subsets of S ;

H_o is a subfamily of H such that

(i) $\alpha, \beta \in H_o \Rightarrow \alpha \cup \beta \in H_o$

(ii) $\alpha \in H$, $\beta \in H_o$ and $\alpha \subset \beta \Rightarrow \alpha \in H_o$;

X is a commutative, Hausdorff semigroup;

$\mu : H \to X$ is finitely additive.

1.1. Definitions.

(1) μ is <u>absolutely continuous relative to</u> H_o , $\mu << (H_o)$, iff, for every $\alpha \in H_o$, $\mu(\alpha) = \underline{0}$.

(2) μ is <u>completely discontinuous (singular) relative to</u> H_o , $\mu \perp (H_o)$, iff, for every $A \in H$ with $\mu(A) \neq \underline{0}$, there exists $\alpha \in H_o$ such that $\alpha \subset A$ and $\mu(\alpha) \neq \underline{0}$.

(3) When λ is a semigroup-valued, finitely additive function on H, the concepts of $\mu \ll \lambda$ and $\mu \perp \lambda$ are obtained from the above by taking

$$H_o = \{\alpha \in H : \lambda(\beta) = \underline{0} \text{ for every } \beta \in H , \beta \subset \alpha\} .$$

(4) When $\overline{\lambda}$ and $\overline{\mu}$ are outer measures on S, the concepts of $\overline{\mu} \ll \overline{\lambda}$ and $\overline{\mu} \perp \overline{\lambda}$ are obtained from the above by taking $H = \mathfrak{M}_{\overline{\lambda}} \cap \mathfrak{M}_{\overline{\mu}}$, $H_o = H \cap \mathfrak{N}_{\overline{\lambda}}$, and $\mu = \overline{\mu}/H$.

1.2. <u>Theorem</u> (Lebesgue Decomposition).

For $A \in H$, let

$$\mu'(A) = \text{limit } \mu(A \setminus \alpha)$$

$$\mu''(A) = \text{limit } \mu(A \cap \alpha)$$

as α runs over H_o directed by $\subset$.

(1) Suppose $\mu'(A)$ and $\mu''(A)$ exist for every $A \in H$. Then

(a) $\mu = \mu' + \mu''$, $\mu' \ll (H_o)$, and $\mu'' \perp (H_o)$;

(b) μ' and μ'' are finitely additive;

(c) if $(H_o)_\sigma = H_o$, X is a uniform semigroup, and μ is σ-additive then μ' and μ'' are also σ-additive.

(2) Suppose μ is s-bounded. Then $\mu'(A)$ and $\mu''(A)$ exist for every $A \in H$ provided

either (a) X is a complete uniform semigroup

or (b) $(H_o)_\sigma = H_o$ and there is a countable base for the neighborhoods of $\underline{0}$ in X. In this case, there exists $\beta \in H_o$ such that, for every $A \in H$,

$$\mu'(A) = \mu(A \setminus \beta) \quad \text{and} \quad \mu''(A) = \mu(A \cap \beta) .$$

Proof:

(1a, b) Immediate from the definitions.

(1c) Let $B, A_n \in H$ for $n \in \omega$, $A_m \cap A_n = \emptyset$ for $m \neq n$ and $B = \bigcup_{n\in\omega} A_n$. Given an element U of the uniformity on X, first choose symmetric elements V_n so that, for any $K \in \omega$,

$$(x_n, y_n) \in V_n \text{ for } n = 0, \cdots, K \quad \Rightarrow \quad (\sum_{n=0}^{K} x_n, \sum_{n=0}^{K} y_n) \in U .$$

Next choose $\beta, \alpha_n \in H_o$ so that, for any $\gamma \in H_o$,

$$\beta \subset \gamma \quad \Rightarrow \quad (\mu'(B), \mu(B \setminus \gamma)) \in U$$

$$\alpha_n \subset \gamma \quad \Rightarrow \quad (\mu'(A_n), \mu(A_n \setminus \gamma)) \in V_n .$$

Let $\gamma = \beta \cup \bigcup_{n\in\omega} \alpha_n$. Then $\gamma \in H_o$,

$$(\mu'(B), \mu(B \setminus \gamma)) \in U \quad \text{and}$$

$$(\mu'(A_n), \mu(A_n \setminus \gamma)) \in V_n \quad \text{for } n \in \omega$$

hence, for any $K \varepsilon \omega$,

$$\left(\sum_{n=0}^{K} \mu(A_n \setminus \gamma),\ \sum_{n=0}^{K} \mu'(A_n)\right) \varepsilon\ U \ .$$

But, for sufficiently large K ,

$$\left(\mu(B \setminus \gamma),\ \sum_{n=0}^{K} \mu(A_n \setminus \gamma)\right) \varepsilon\ U$$

and therefore

$$\left(\mu'(B),\ \sum_{n=0}^{K} \mu'(A_n)\right) \varepsilon\ U \circ U \circ U \ .$$

Thus, μ' is σ-additive. Similarly for μ'' .

(2) Since μ is s-bounded, for any neighborhood U of $\underline{0}$ in X , there exists $\alpha \varepsilon H_o$ such that, for any $\gamma \varepsilon H_o$,

$$\alpha \cap \gamma = \emptyset \quad \Rightarrow \quad \mu(\gamma) \varepsilon\ U$$

hence, for any $A \varepsilon H$ and $\beta \varepsilon H_o$,

$$\begin{aligned} \alpha \subset \beta \quad \Rightarrow \quad \mu(A \setminus \alpha) &= \mu(A \setminus \beta) + \mu(A \cap \beta \setminus \alpha) \\ &\varepsilon\ \mu(A \setminus \beta) + U \ . \end{aligned}$$

(2a) Thus, the $\mu(A \setminus \alpha)$ form a Cauchy net as α runs over H_o so that, since X is complete $\mu'(A)$ exists. Similarly for $\mu''(A)$.

(2b) If $\{U_n; n \in \omega\}$ is a countable base for the neighborhoods of $\underline{0}$ in X, choose $\alpha_n \in H_o$ so that, for any $\gamma \in H_o$,

$$\alpha_n \cap \gamma = \emptyset \quad \Rightarrow \quad \mu(\gamma) \in U_n$$

and let $\beta = \bigcup_{n \in \omega} \alpha_n$. Then $\beta \in H_o$ and, for every $\gamma \in H_o$,

$$\beta \cap \gamma = \emptyset \quad \Rightarrow \quad \mu(\gamma) = \underline{0}\ .$$

Therefore, for every $A \in H$ and $\gamma \in H_o$,

$$\beta \subset \gamma \quad \Rightarrow \quad \mu(A \setminus \beta) = \mu(A \setminus \gamma) + \mu(A \cap \gamma \setminus \beta) = \mu(A \setminus \gamma)$$

$$\text{and} \quad \mu(A \cap \gamma) = \mu(A \cap \beta) + \mu(A \cap \gamma \setminus \beta) = \mu(A \cap \beta)$$

Thus, $\mu'(A) = \mu(A \setminus \beta)$ and $\mu''(A) = \mu(A \cap \beta)$. I

Remark. Note that, in 2a the condition that X be a complete uniform semigroup could be replaced by the weaker hypothesis that the range of μ be s-precomplete.

2. Completion.

Starting with two measures λ, μ on a common family H subject to the condition $\mu << \lambda$, we want to consider extensions to outer measures $\overline{\lambda}$, $\overline{\mu}$ which also satisfy the condition $\overline{\mu} << \overline{\lambda}$.

Throughout this section we suppose

S is an abstract space and

H is a ring of subsets of S.

2.1. Definitions.

For any semigroup-valued function λ on H,

(1) $H_o(\lambda) = \{\alpha \in H : \lambda(\beta) = \underline{0}$ for every $\beta \in H, \beta \subset \alpha\}$

(2) $\text{Null } \lambda = \{A : A \subset \alpha$ for some $\alpha \in H_o(\lambda)\}$

(3) $\overline{H}_\lambda = \{A : ((A \setminus \alpha) \cup (\alpha \setminus A)) \in \text{Null } \lambda$ for some $\alpha \in H\}$

$\overline{H}_\lambda$ is the <u>completion</u> of H relative to λ.

<u>Remark.</u> If λ is finitely additive then $\overline{H}_\lambda$ is also a ring and there is a unique trivial finitely additive extension λ' of λ to $\overline{H}_\lambda$ such that $\lambda'(A) = \underline{0}$ for $A \in \text{Null } \lambda$. If H is a σ-ring and λ is σ-additive then $\overline{H}_\lambda$ is also a σ-ring and λ' is σ-additive.

2.2. Definition.

For any function ψ on sb S and $F \subset$ sb S , A' is a hull of A in (F, ψ) iff $A \subset A' \varepsilon F$ and, for every E ,

$$A \subset E \subset A' \quad \Rightarrow \quad \psi(A) = \psi(E) .$$

Remark. Recall definition 1.1(2) of U-hull in Chapter I. When range ψ is Hausdorff, A' is a hull of A in (F, ψ) iff, for every neighborhood U of $\psi(A)$, A' is a U-hull of A in (F, ψ) .

2.3. Lemma. If ψ is an outer measure with values in a commutative, Hausdorff group and $F \subset$ sb S then A' is a hull of A in (F, ψ) iff $A \subset A' \varepsilon F$ and, for every $\alpha \varepsilon \mathfrak{M}_\psi$,

$$\psi(A \cap \alpha) = \psi(A' \cap \alpha) .$$

Proof: Suppose A' is a hull of A and $\alpha \varepsilon \mathfrak{M}_\psi$. Let $E = (A \setminus \alpha) \cup (A' \cap \alpha)$. Then $A \subset E \subset A'$ so $\psi(A) = \psi(E)$. But

$$\psi(A) = \psi(A \cap \alpha) + \psi(A \setminus \alpha)$$

$$\psi(E) = \psi(E \cap \alpha) + \psi(E \setminus \alpha) = \psi(A' \cap \alpha) + \psi(A \setminus \alpha) .$$

Therefore $\psi(A \cap \alpha) = \psi(A' \cap \alpha)$.

Suppose now that $A \subset A' \in F$ and, for every $\alpha \in \mathfrak{M}_\psi$, $\psi(A \cap \alpha) = \psi(A' \cap \alpha)$. Given $A \subset E \subset A'$ and neighborhood U of $\psi(E)$, in view of theorem 3.2(iii) in Chapter I, there exists $\alpha \in \mathfrak{M}_\psi$ such that α is a U-hull of E . Then

$$\psi(A) = \psi(A \cap \alpha) = \psi(A' \cap \alpha) \in U \ .$$

Therefore, $\psi(A) = \psi(E)$. I

We now state the main result we shall need in subsequent sections.

2.4. <u>Theorem.</u> Suppose H is a σ-field and λ , μ are σ-additive, semigroup-valued functions on H with $\mu \ll \lambda$. If λ takes values in a metric group then there exist H-outer measures $\bar{\lambda}$, $\bar{\mu}$ on S such that

(1) $\bar{\lambda}/H = \lambda$ and $\bar{\mu}/H = \mu$;

(2) for every $A \subset S$, there is an A' which is a hull of A in $(H, \bar{\lambda})$ and in $(H, \bar{\mu})$;

(3) $\mathfrak{N}_{\bar{\lambda}}$ = Null λ and $\mathfrak{M}_{\bar{\lambda}} = \bar{H}_\lambda$;

(4) $\mathfrak{N}_{\bar{\lambda}} \subset \mathfrak{N}_{\bar{\mu}}$ and $\mathfrak{M}_{\bar{\lambda}} \subset \mathfrak{M}_{\bar{\mu}}$ so that $\bar{\mu} \ll \bar{\lambda}$.

<u>Proof:</u> For any $A \subset S$, let $H^+(A) = \{\alpha \in H : A \subset \alpha\}$ directed by $\supset$ and set

$$\bar{\lambda}(A) = \text{limit } \lambda(\alpha) \quad \text{as } \alpha \text{ runs over } H^+(A)$$

$$\bar{\mu}(A) = \text{limit } \mu(\alpha) \quad \text{as } \alpha \text{ runs over } H^+(A) \; .$$

Since there is a countable base for the neighborhoods of $\underline{0}$ in the range of λ and, for any descending sequence α in H, $\lim_n \lambda(\alpha_n) = \lambda(\bigcap_{n\varepsilon\omega} \alpha_n)$, we see that there exists $A' \varepsilon H^+(A)$ such that $\bar{\lambda}(A) = \lambda(A')$ and in fact, for any $\alpha \varepsilon H$,

$$\begin{aligned} A \subset \alpha \subset A' \quad &\Rightarrow \quad \lambda(\alpha) = \lambda(A') \\ &\Rightarrow \quad (A' \setminus \alpha) \varepsilon H_o(\lambda) \subset H_o(\mu) \\ &\Rightarrow \quad \mu(\alpha) = \mu(A') \; . \end{aligned}$$

Thus, $\bar{\mu}(A) = \mu(A')$ and A' is a hull of A in $(H, \bar{\lambda})$ and in $(H, \bar{\mu})$. If $A \varepsilon \mathcal{N}_{\bar{\lambda}}$ then $A' \varepsilon H_o(\lambda)$ hence $A \varepsilon \text{Null } \lambda$. On the other hand, we clearly have $\text{Null } \lambda \subset \mathcal{N}_{\bar{\lambda}}$. Thus, $\mathcal{N}_{\bar{\lambda}} = \text{Null } \lambda$. Since $H_o(\lambda) \subset H_o(\mu)$, we have $\text{Null } \lambda \subset \text{Null } \mu \subset \mathcal{N}_{\bar{\mu}}$ and hence $\mathcal{N}_{\bar{\lambda}} \subset \mathcal{N}_{\bar{\mu}}$. Finally, if $A \varepsilon \mathcal{M}_{\bar{\lambda}}$ then $(A' \setminus A) \varepsilon \mathcal{N}_{\bar{\lambda}}$ and therefore $A \varepsilon \bar{H}_\lambda$. Thus, $\mathcal{M}_{\bar{\lambda}} = \bar{H}_\lambda$. Since $\mathcal{N}_{\bar{\lambda}} \subset \mathcal{N}_{\bar{\mu}} \subset \mathcal{M}_{\bar{\mu}}$, we conclude $\bar{H}_\lambda \subset \mathcal{M}_{\bar{\mu}}$. I

3. Differentiation Bases.

In the classical theory of differentiation in Euclidean space, for the derivative of μ with respect to λ at a point s one takes the limit of ratios $\mu(\alpha)/\lambda(\alpha)$ as α runs over the family of spheres (or cubes) containing s and the diameter of α tends to zero. A key property passessed by spheres in relation to Lebesgue measure λ is the so called Vitali Property: if every point of a set A can be covered by a sphere in a given family F having arbitrarily small diameter then a countable, disjoint family of spheres in F will cover λ-almost all of A. This property plays an essential role in the development of the theory.

When working in an abstract space, the search for a suitable substitute for spheres, incorporating the idea of "shrinking to a point" and possessing Vitali-like properties, leads to consideration of various types of differentiation bases.

3.1. Definitions.

(1) $\mathfrak{F}$ is a <u>differentiation basis</u> on S iff, for every $s \varepsilon S$, $\mathfrak{F}(s)$ is a filterbase of families of subsets of S.

(2) For a differentiation basis $\mathfrak{F}$,

$$R(\mathfrak{F}) = \{\alpha : \text{for some } s \varepsilon S \text{ and } F \varepsilon \mathfrak{F}(s),\ \alpha \varepsilon F\} .$$

<u>Note</u>: To appreciate the above definitions, it may help to keep in mind that an $F \in \mathcal{F}(s)$ corresponds to an $\varepsilon > 0$ in the classical situation, an $\alpha \in F$ to a sphere of diameter less than ε about s, and $R(\mathcal{F})$ to the family of all spheres. The notion of limit as α runs over $\mathcal{F}(s)$ corresponds to that of limit as "α shrinks to s."

3.2. <u>Definitions</u>.

Let $\mathcal{F}$ be a differentiation basis on S.

(1) For $A \subset S$ and $G \subset R(\mathcal{F})$,
G is a <u>Vitali cover</u> for A iff, for every $s \in A$ and $F \in \mathcal{F}(s)$, $G \cap F \neq \emptyset$.

(2) For any outer measure λ on S,
$\mathcal{F}$ is a <u>Vitali system</u> for λ iff $R(\mathcal{F}) \subset \mathfrak{M}_\lambda$ and, for every $A \subset S$, neighborhood U of $\lambda(A)$ and Vitali cover G for A, there exists a countable, disjoint $G' \subset G$ such that if $A' = \bigcup_{\alpha \in G'} \alpha$ then

(i) $(A \setminus A') \in \mathfrak{N}_\lambda$ and

(ii) A' is a U-hull of $(A \cap A')$ in $(\mathfrak{M}_\lambda, \lambda)$, i.e., for every E,

$$A \cap A' \subset E \subset A' \quad \Rightarrow \quad \lambda(E) \in U .$$

(3) For any outer measures λ , μ on S , $\mathcal{J}$ is a joint tile system for λ , μ iff $R(\mathcal{J}) \subset \mathcal{M}_\lambda \cap \mathcal{M}_\mu$ and, for every $A \subset S$, neighborhood U of $\lambda(A)$ and neighborhood V of $\mu(A)$, and Vitali cover G for A , there exists a countable $G' \subset G$ such that

$$\sum_{\alpha \in G'} \lambda(\alpha) \in U \quad \text{and} \quad \sum_{\alpha \in G'} \mu(\alpha) \in V \ .$$

Examples.

(1) Classical Vitali system used for ordinary differentiation.

Let $S = \mathcal{R}$ and, for any $s \in S$ and $\varepsilon > 0$,

$$F(s, \varepsilon) = \{\alpha : \alpha \text{ is a closed, bounded interval with } s \text{ as an end-point and } 0 < \operatorname{diam} \alpha < \varepsilon\} ,$$

$$\mathcal{J}(s) = \{F(s, \varepsilon) \ ; \ \varepsilon > 0\} \ .$$

Then $\mathcal{J}$ is a Vitali system for Lebesgue outer measure.

(2) Sequence of partitions.

For $n \in \omega$, let P_n be a countable partition of S with P_{n+1} a refinement of P_n and, for $s \in S$, let $\alpha_n(s)$ be the $\alpha \in P_n$ with $s \in \alpha$. Let

$$F_n(s) = \{\alpha_k(s) \ ; \ k \geq n\} \ ,$$

$$\mathcal{J}(s) = \{F_n(s) \ ; \ n \in \omega\} ,$$

$$H = \mathcal{R}(\mathcal{J}) = \bigcup_{n\in\omega} P_n \ .$$

Then $\mathcal{J}$ is a Vitali system for any H-outer measure on S .

(3) Separable metric space.

Let S be a separable metric space and, for each $n \in \omega_+$, let G_n be a countable cover of S by open sets of diameter less than $\frac{1}{n}$ and Q_n be the partition obtained by well ordering G_n and subtracting from each element in G_n all the preceding ones. Define P_n by recursion so that $P_1 = Q_1$ and

$$P_{n+1} = \{\alpha \cap \beta \ ; \ \alpha \in P_n \text{ and } \beta \in Q_{n+1}\} \ .$$

Then P_n is a partition of S , P_{n+1} is a refinement of P_n and $\sup\{\operatorname{diam} \alpha \ ; \ \alpha \in P_n\} \leq \frac{1}{n}$. Let $\mathcal{J}$ be the differentiation basis obtained from $\{P_n \ ; \ n \in \omega\}$ as in example 2. Then $\mathcal{J}$ is a Vitali system for any $\mathcal{G}$-outer measure on S , where $\mathcal{G}$ is the family of open sets.

Note: It is known that if $F(n, s)$ is the family of all open rectangles in the plane centered at s and of diameter less than $\frac{1}{n}$ and $\mathcal{J}(s) = \{F(n, s) \ ; \ n \in \omega_+\}$ then $\mathcal{J}$ is not a Vitali system for Lebesgue outer measure on the plane.

When S is an abstract space and λ is a real valued outer measure on S, the existence of a Vitali system for λ is guaranteed by the Lifting Theorem in conjunction with the following.

3.3. Theorem. Let λ be an outer measure on S and, for $A, B \in \mathfrak{M}_\lambda$, Let

$$A \equiv B \iff ((A \setminus B) \cup (B \setminus A)) \in \mathfrak{N}_\lambda .$$

Let $\rho : \mathfrak{M}_\lambda \to \mathfrak{M}_\lambda$ be such that, for $A, B \in \mathfrak{M}_\lambda$,

(i) $A \equiv \rho(A)$

(ii) $A \equiv B \implies \rho(A) = \rho(B)$

(iii) $\rho(A \cap B) = \rho(A) \cap \rho(B)$

(iv) $\rho(S) = S$ and $\rho(\emptyset) = \emptyset$.

Let $R = \text{range}\, \rho \setminus \{\emptyset\}$ and, for every $s \in S$, $\mathfrak{F}(s)$ be the filterbase associated with $\{\alpha \in R : s \in \alpha\}$ directed by $\supset$, i.e.,

$$\mathfrak{F}(s) = \{F : F = \{\alpha \in R : s \in \alpha \subset \beta\} \text{ for some } \beta \in R, s \in \beta\}.$$

If every disjoint family in R is countable then $\mathfrak{F}$ is a Vitali system for λ.

[Note: The role of the Lifting Theorem is to produce such a ρ when λ is real valued.]

Proof:

(i) Given $B \subset S$ and a Vitali cover F for B, let F' be a maximal, disjoint subfamily of F and $A' = \bigcup_{\alpha \in F'} \alpha$.

Then F' is countable and $A' \in \mathcal{M}_\lambda$. We shall check that $(B \setminus A') \in \mathcal{N}_\lambda$. To this end, let $\beta = \rho(S \setminus A')$. We must have $B \cap \beta = \emptyset$, for if $s \in B \cap \beta$ then, since F is a Vitali cover for B, there exists $\alpha \in F$ with $s \in \alpha \subset \beta$ so that, for every $\alpha' \in F'$,

$$\alpha \cap \alpha' \subset \beta \cap \alpha' \equiv (S \setminus A') \cap \alpha' = \emptyset$$

and therefore

$$\alpha \cap \alpha' = \rho(\alpha) \cap \rho(\alpha') = \rho(\alpha \cap \alpha') = \emptyset \ .$$

This, however, contradicts the maximality of F'. Thus,

$$(B \setminus A') \equiv B \cap \beta = \emptyset \quad \text{so} \quad (B \setminus A') \in \mathcal{N}_\lambda \ .$$

(ii) Given $A \subset S$, a neighborhood U of $\lambda(A)$ and a Vitali cover G for A, let B' be a U-hull of A in $(\mathcal{M}_\lambda, \lambda)$, $B = A \cap \rho(B')$ and

$$F = \{\alpha \in G : \alpha \subset \rho(B')\} \ .$$

Then

$$(A \setminus B) \subset B' \setminus \rho(B') \in \mathcal{N}_\lambda$$

and F is a Vitali cover for B. Taking F' and A' as in (i), we see that

$$(A \setminus A') \subset (A \setminus B) \cup (B \setminus A') \in \mathcal{N}_\lambda$$

and A' is a U-hull of $(A \cap A')$ in $(\mathcal{N}_\lambda, \lambda)$ since, for any E,

$$A \cap A' \subset E \subset A' \quad \Rightarrow \quad (E \setminus B') \subset (\rho(B') \setminus B') \in \mathcal{N}_\lambda$$

$$\text{and} \quad (A \setminus E) = (A \setminus A') \in \mathcal{N}_\lambda$$

$$\Rightarrow \quad \lambda(E) = \lambda(A \cup (E \cap B')) \in U . \;\blacksquare$$

We conclude this section with a theorem which relates Vitali systems to joint tile systems in the presence of absolute continuity.

3.4. Theorem. Let λ, μ be H-outer measures on S for some family H and $\mu << \lambda$. If λ takes values in a metric group and μ is s-bounded then any Vitali system $\mathcal{J}$ for λ with $\mathcal{R}(\mathcal{J}) \subset H$ is a joint tile system for λ, μ.

Proof: We first check the following.

If $\alpha_{n+1} \subset \alpha_n \in \mathcal{M}_\mu$ for $n \in \omega$ and $\beta = \bigcap_{n\in\omega} \alpha_n$ then, for any neighborhood V of $\mu(\beta)$, there exists

$n \in \omega$ such that α_n is a V-hull of β in $(\mathcal{M}_\mu, \mu)$.

Indeed, let V_o, V_1 be neighborhoods of $\underline{0}$ in range μ with closure $V_1 \subset V_o$ and $\mu(\beta) + V_o \subset V$. If, for every $n \in \omega$, $(\alpha_n \setminus \beta)$ is not a V_o-hull of $\emptyset$ then there exists $\gamma_n \subset (\alpha_n \setminus \beta)$ with $\mu(\gamma_n) \notin V_o$ and therefore, for k large enough, $\mu(\gamma_n \cap (\alpha_n \setminus \alpha_k)) \notin V_1$. Therefore we can choose $k_n \in \omega$ and $\gamma_n \in \mathcal{M}_\mu$ by recursion so that $k_n < k_{n+1}$, $\gamma_n \subset (\alpha_{k_n} \setminus \alpha_{k_{n+1}})$ and $\mu(\gamma_n) \notin V_1$. Since the γ_n are disjoint, this contradicts the fact that μ is s-bounded.

Now, let $\overline{H} = \mathcal{M}_\lambda \cap \mathcal{M}_\mu$ and $\{U_n ; n \in \omega\}$ be a base for the neighborhoods of $\underline{0}$ in range λ with the property

$$\sum_{n>N} U_n \subset U_N \quad \text{for any} \quad N \in \omega .$$

Given $A \subset S$ and a Vitali cover G for A, choose a countable, disjoint $F_n \subset G$ and $B_n = \bigcup_{\alpha \in F_n} \alpha$ so that $(A \setminus B_n) \in \mathcal{N}_\lambda$ and B_n is a U_n-hull of $(A \cap B_n)$ in $(\overline{H}, \lambda)$. Let

$$B' = \bigcap_{N \in \omega} \bigcup_{n>N} B_n$$

and

$$A_o = A \cap \bigcap_{n \in \omega} B_n .$$

Then $(A \setminus A_o) \in \mathcal{N}_\lambda \subset \mathcal{N}_\mu$. We shall check that B' is a hull of A_o in $(\overline{H}, \lambda)$ and $(\overline{H}, \mu)$. Indeed, for any $\beta \in \mathcal{M}_\lambda$ with $A_o \subset \beta$ and $\alpha \subset (B_n \setminus \beta)$, we have $\lambda(\alpha) \in U_n$ hence

$$\alpha \subset (\bigcup_{n>N} B_n \setminus \beta) \quad \Rightarrow \quad \lambda(\alpha) \in \sum_{n>N} U_n \subset U_N$$

and therefore $(B' \setminus \beta) \in \mathcal{N}_\lambda \subset \mathcal{N}_\mu$. We conclude that B' is a hull of A_o in $(\overline{H}, \lambda)$ and $(\overline{H}, \mu)$. Given any neighborhoods U of $\lambda(A)$ and V of $\mu(A) = \mu(A_o) = \mu(B')$, letting $\alpha_N = \bigcup_{n>N} B_n$ in (i), we see that for some $N \in \omega$, α_N is a V-hull of B' and hence, for any $n > N$, B_n is a V-hull of A_o . Then for $n > N$ with $\lambda(A) + U_n \subset U$ we have

$$\sum_{\alpha \in F_n} \lambda(\alpha) \in U \quad \text{and} \quad \sum_{\alpha \in F_n} \mu(\alpha) \in V \ .$$

I

4. The Derivative.

Throughout this section we suppose

S is an abstract space;

H is a σ-field of subsets of S ;

$\mathfrak{F}$ is a differentiation basis with $R(\mathfrak{F}) \subset H$;

$\lambda : H \to [0, \infty)$ is σ-additive and $\lambda(\alpha) > 0$ for $\alpha \in R(\mathfrak{F})$;

X is a topological, Hausdorff, vector space;

$\mu : H \to X$ is σ-additive.

We are interested in studying the limit of $\mu(\alpha)/\lambda(\alpha)$ as α runs over $\mathfrak{F}(s)$ for $s \in S$. Since this limit may not exist, we consider instead all possible limit points of the ratios, i.e. we let

$$D(s) = \bigcap_{F \in \mathfrak{F}(s)} \text{closure}\ \{\mu(\alpha)/\lambda(\alpha)\ ;\ \alpha \in F\} \ .$$

We refer to D as the "outer derivative of μ w.r.t. λ ." (This is a substitute for the notions of lim inf and lim sup of the ratios in the classical case). The problem in general is to determine when $D(s)$ is not empty, when does it consist of a single point (in which case, we refer to $D(s)$ as the ordinary derivative), and for which $A \subset S$ do we have

$$\mu(A) = \int_A D\, d\lambda \ .$$

Our main result is the following.

4.1. Fundamental Theorem.

Let $\mu \ll \lambda$ and $\bar{\lambda}$, $\bar{\mu}$ be the H-outer measures generated by λ , μ respectively (theorem 2.4). If

(i) X is locally convex,

(ii) $\mathcal{J}$ is a joint tile system for $\bar{\lambda}$, $\bar{\mu}$,

(iii) $D(s) \neq \emptyset$ for $\bar{\lambda}$-almost all $s \in S$,

(iv) D is $\bar{\lambda}$-quasi bounded

then

(1) for any neighborhood U of $\underline{0}$ in X ,

$$D(s) - D(s) \subset U \qquad \text{for } \bar{\lambda}\text{-almost all } s \in S ,$$

(2) D is $\bar{\lambda}$-partitionable,

(3) $\bar{\mu}(A) = \int_A D \, d\lambda$ for any $A \subset S$.

We shall give the proof in a series of lemmas all of which assume the hypotheses of the theorem.

Lemma 1. Let $A \subset S$, $\bar{\lambda}(A) > 0$ and W be a convex subset of X . If, for every $s \in A$ and $F \in \mathcal{J}(s)$, there exists $\alpha \in F$ with $(\mu(\alpha)/\lambda(\alpha)) \in W$ then $(\bar{\mu}(A)/\bar{\lambda}(A)) \in$ closure W .

<u>Proof:</u> Let

$$G = \{\alpha \in R(\mathcal{J}) : \mu(\alpha)/\lambda(\alpha) \in W\} .$$

Then G is a Vitali cover for A and so, for any $\varepsilon > 0$ and symmetric neighborhood U of $\underline{0}$ in X, there is a countable, disjoint $G' \subset G$ such that

$$\left| \sum_{\alpha \in G'} \lambda(\alpha) - \overline{\lambda}(A) \right| < \varepsilon$$

and

$$\left(\sum_{\alpha \in G'} \mu(\alpha) - \overline{\mu}(A) \right) \in \overline{\lambda}(A) \cdot U .$$

Since W is convex, for any finite $\Delta \subset G'$, letting $a = \sum_{\alpha \in \Delta} \lambda(\alpha)$ and $b = \sum_{\alpha \in \Delta} \mu(\alpha)$, we have

$$\frac{b}{a} = \sum_{\alpha \in \Delta} \frac{\lambda(\alpha)}{a} \cdot \left(\frac{\mu(\alpha)}{\lambda(\alpha)} \right) \in W .$$

Therefore, passing to the limit as Δ increases, we get

$$\left(\sum_{\alpha \in G'} \mu(\alpha) / \sum_{\alpha \in G'} \lambda(\alpha) \right) \in \text{closure } W .$$

Since ε and U are arbitrary, we conclude $(\overline{\mu}(A)/\overline{\lambda}(A)) \in$ closure W. ∎

<u>Lemma 2.</u> Let U be a convex, open neighborhood of $\underline{0}$ in X and U' be a translate of U. If

$$A = \{s \in S : D(s) \cap U' \neq \emptyset\}$$

then there exists $A' \in H$ such that $A \subset A'$ and, for $\overline{\lambda}$-almost all $s \in A'$,

$$D(s) \subset U' + U \quad .$$

<u>Proof:</u> By theorem 2.4, let A' be a hull of A in $(H, \overline{\lambda})$ and in $(H, \overline{\mu})$. Then, by lemma 2.3, for every $\alpha \in \mathfrak{M}_{\overline{\lambda}} \subset \mathfrak{M}_{\overline{\mu}}$, we have

$$\overline{\lambda}(A \cap \alpha) = \overline{\lambda}(A' \cap \alpha) \qquad \text{and}$$

$$\overline{\mu}(A \cap \alpha) = \overline{\mu}(A' \cap \alpha) \quad .$$

Let V be a symmetric, convex, open neighborhood of $\underline{0}$ in X with $4V \subset U$, V' be a translate of V with closure $V' \cap$ closure $U' = \emptyset$ and

$$B = \{s \in A' : D(s) \cap V' \neq \emptyset\} \quad .$$

We first check that $\overline{\lambda}(B) = 0$. Indeed, if $\overline{\lambda}(B) > 0$, let B' be a hull of B in $(H, \overline{\lambda})$ and in $(H, \overline{\mu})$. Then

$$\overline{\lambda}(B) = \overline{\lambda}(B \cap A') = \overline{\lambda}(B' \cap A') = \overline{\lambda}(B' \cap A)$$

$\overline{\mu}(B) = \overline{\mu}(B \cap A') = \overline{\mu}(B' \cap A') = \overline{\mu}(B' \cap A)$.

But, by lemma 1,

$$(\overline{\mu}(B)/\overline{\lambda}(B)) \in \text{closure } V'$$

$$(\overline{\mu}(B' \cap A)/\overline{\lambda}(B' \cap A)) \in \text{closure } U' .$$

This contradicts the choice of V' . Thus $\overline{\lambda}(B) = 0$.

Now, since D is $\overline{\lambda}$-quasi bounded, there exist $S' \subset S$ with $\overline{\lambda}(S') = 0$ and a countable family J of translates of V which covers $D[S \setminus S']$. Let

$$S'' = \{s \in A' : D(s) \cap V' = \emptyset \text{ for some } V' \in J \text{ with } \text{closure } V' \cap \text{closure } U' = \emptyset\} .$$

Then $\overline{\lambda}(S'') = 0$ and, for every $s \in (A' \setminus (S' \cup S''))$, $D(s) \subset \text{closure } U' + 3V \subset U' + U$. ∎

<u>Lemma 3.</u> D is H-partitionable.

<u>Proof:</u> Given a neighborhood U of $\underline{0}$ in X , let V be a convex, open neighborhood of $\underline{0}$ with $4V \subset U$ then choose $S' \subset S$ with $\overline{\lambda}(S') = 0$ and a countable family J of translates of V which covers $D[S \setminus S']$. For each $W \in J$, let

$$A_W = \{s \in S : D(s) \cap W \neq \emptyset\}$$

and, by lemma 2, let $B_W \varepsilon H$ be such that, for $\bar{\lambda}$-almost all $s \varepsilon B_W$, $D(s) \subset W + V$. Let

$$S_o = S' \cup \bigcup_{W \varepsilon J} \{s \varepsilon B_W : D(s) \not\subset W + V\} \quad .$$

Then $\bar{\lambda}(S_o) = 0$ and, for any $s,t \varepsilon (B_W \setminus S_o)$, $D(s) \subset W + V$ and $D(t) \subset W + V$ hence

$$D(s) - D(t) \subset 4V \subset U$$

so $D[B_W \setminus S_o]$ is contained in a translate of U. ▌

Lemma 4. For any $A \subset S$,

$$\bar{\mu}(A) = \int_A D \, d\lambda \quad .$$

Proof: Given any convex neighborhood U of $\underline{0}$ in X, by lemma 3, there exist $S' \subset S$ with $\lambda(S') = 0$ and a countable partition P of $(A \setminus S')$ by elements of H such that, for every $\alpha \varepsilon P$, $D[\alpha] - D[\alpha] \subset U$. Given $A \varepsilon H$, let Q be any countable partition of $A \setminus S'$ by elements of H which is a refinement of P. For any $\alpha \varepsilon Q$ with $\lambda(\alpha) > 0$, we must have

$$D[\alpha] \subset (\mu(\alpha)/\lambda(\alpha)) + 3U$$

since there exists $a \varepsilon X$ such that, for every $s \varepsilon \alpha$,

$D(s) \subset a + U$ and hence, by lemma 1,

$$(\mu(\alpha)/\lambda(\alpha)) \in \text{closure } (a + U)$$

so $\quad a \in (\mu(\alpha)/\lambda(\alpha)) + 2U$.

Therefore,

$$\sum_{\alpha \in Q} D[\alpha] \cdot \lambda(\alpha) \subset \sum_{\alpha \in Q} (\mu(\alpha)/\lambda(\alpha) + 3U) \cdot \lambda(\alpha)$$

$$\subset \mu(A) + 3U \cdot \lambda(S) \quad .$$

Since $D[\alpha] \neq \emptyset$ for some $\alpha \in Q$, we conclude

$$\int_A D \, d\lambda = \mu(A) \quad .$$

Since $\overline{\mu}$ is an H-outer measure, we must have, for any $A \subset S$,

$$\overline{\mu}(A) = \int_A D \, d\lambda = \int_A D \, d\overline{\lambda} \quad .$$

I

<u>Remarks</u>.

(a) In general, D may be multiple-valued but its integral is always a point in X . When there is a countable base for the neighborhoods of $\underline{0}$ in X then D is single-valued almost everywhere. Even when this does not occur, if f is <u>any</u> function with $f(s) \in D(s)$ for $\overline{\lambda}$-almost all $s \in S$ then, by the definition of integral,

$$\bar{\mu}(A) = \int_A f \, d\lambda \qquad \text{for} \quad A \subset S .$$

Thus, the above theorem guarantees that $\bar{\mu}$ has an integral representation with a single-valued function as integrand.

(b) In view of the lifting theorem and theorems 3.3 and 3.4, a joint tile system for $\bar{\lambda}$, $\bar{\mu}$ always exists. In particular applications, however, the use of some special system frequently yields special properties of the derivative D and hence a stronger integral representation theorem.

(c) Hypothesis (iii) that $D(s) \neq \emptyset$ is usually satisfied by imposing some kind of compactness condition. Clearly, if the ratios $\mu(\alpha)/\lambda(\alpha)$ are in a compact set for $\alpha \in F \in \mathfrak{F}(s)$ then $D(s) \neq \emptyset$. The fact that the unit ball in the dual of a Banach space is compact in the weak* topology explains the role of such spaces in many integral representation theorems.

(d) Hypothesis (iv) that D be $\bar{\lambda}$-quasi bounded is usually satisfied by rather mild countability conditions on the topology of X. It is automatically satisfied when X is endowed with the weak topology, since a countable number of translates of a weak neighborhood of $\underline{0}$ cover all of X. The hypothesis is also satisfied when X is Lindelöf. Frequently, the compactness condition imposed to yield (iii) will also yield (iv).

In the light of the above remarks, theorem 4.1 yields as a corollary the following general representation theorem.

4.2. <u>Theorem</u>. Let $H = \overline{H}_\lambda$ (i.e., H be complete relative to λ), $H^+ = \{\alpha \in H : \lambda(\alpha) > 0\}$, and $\mu << \lambda$. If

(i) X is locally convex and

(ii) for any $\beta' \in H^+$ there exist $\beta \in H^+$ with $\beta \subset \beta'$ and compact $C \subset X$ such that

$$\{\mu(\alpha)/\lambda(\alpha) ; \alpha \in H^+ \text{ and } \alpha \subset \beta\} \subset C$$

then there is an H-partitionable, single-valued function f such that

$$\mu(A) = \int_A f \cdot d\lambda \qquad \text{for } A \in H .$$

<u>Proof</u>: Let J be a maximal disjoint family of sets $\beta \in H^+$ for which there exists a compact $C \subset X$ with

$$\{\mu(\alpha)/\lambda(\alpha) : \alpha \in H^+ \text{ and } \alpha \subset \beta\} \subset C .$$

Let $S' = \bigcup_{\beta \in J} \beta$. Then J is countable and, in view of condition (ii), $\lambda(S \setminus S') = 0$. We may suppose $S = S'$, for otherwise replace some $\beta \in J$ by $\beta \cup (S \setminus S')$. Let $\overline{\lambda}$, $\overline{\mu}$ be the H-outer measures generated by λ, μ respectively (theorem 2.4); ρ be a lifting of H ($= \mathfrak{M}_{\overline{\lambda}}$) with $\rho(\beta) = \beta$ for $\beta \in J$ (lifting theorem); $\mathfrak{I}$ be the Vitali system for $\overline{\lambda}$ associated with ρ (theorem 3.3). Then $\mathfrak{I}$ is a joint tile system for $\overline{\lambda}$, $\overline{\mu}$ (theorem 3.4) and all the hypotheses of theorem 4.1 are satisfied, the last one because $D[S']$ is covered by a countable family of compact sets. Hence, any function f with $f(s) \in D(s)$ for λ-almost all $s \in S$ has the desired property. ∎

5. Special Integral Representation Theorems.

We shall now obtain a few of the better known integral representation theorems as special cases of the previous results. The basic set up is the same as in section 4. For simplicity, we also assume that H is complete, i.e. $H = \overline{H}_\lambda$ and let $H^+ = \{\alpha \in H : \lambda(\alpha) > 0\}$.

5.1. Theorem. Let X be the dual of a Banach space and be endowed with the weak* topology, $\mu << \lambda$, and $\{\mu(\alpha)/\lambda(\alpha) \; ; \; \alpha \in H^+\}$ be bounded. Then there exists $f : S \to X$ such that

$$\mu(A) = \int_A f \cdot d\lambda \qquad \text{for} \quad A \in H .$$

Proof: Apply theorem 4.2 after noting that a bounded set in X is relatively compact.

As immediate corollaries we obtain the following.

5.2. Theorem. Let X be the dual of a Banach space Y and be endowed with the weak* topology, U be a bounded linear operator on $\mathcal{L}_1(\lambda)$ to X . Then there exists a partitionable (i.e. weak* measurable) $f : S \to X$ such that, for every $\xi \in \mathcal{L}_1(\lambda)$,

$$U(\xi) = \int_S (\xi \cdot f) d\lambda$$

i.e.,

$$U(\xi)(y) = \int_S (\xi(s)\cdot f(s)(y))\,d\lambda(s) \quad \text{for} \quad y \in Y .$$

<u>Proof:</u> For $\alpha \in H$, let $\mu(\alpha) = U(1_\alpha)$. Then $\mu << \lambda$ and $\{\mu(\alpha)/\lambda(\alpha) ; \alpha \in H^+\}$ is bounded so, by theorem 5.1,

$$U(1_\alpha) = \int_\alpha f\cdot d\lambda = \int_S (1_\alpha\cdot f)\,d\lambda \quad \text{for} \quad \alpha \in H .$$

Since simple functions are dense in $\mathcal{L}_1(\lambda)$, we conclude

$$U(\xi) = \int_S (\xi\cdot f)\,d\lambda \quad \text{for} \quad \xi \in \mathcal{L}_1(\lambda) .$$

I

<u>Note:</u> As an immediate consequence of the way f is obtained, we have $\|f(s)\| \leq \|U\|$ for $s \in S$. Hence, in view of the conclusion,

$$\sup_{s\in S} \|f(s)\| = \|U\| .$$

5.3. <u>Theorem.</u> Let T be a locally compact space and X be the space of real-valued Radon outer measures on T endowed with the weak* (or vague) topology. If $\nu \in X$ and $\pi : T \to S$ are such that, for every $\alpha \in H$, $\pi^{-1}[\alpha] \in \mathcal{M}_\nu$ and $\nu(\pi^{-1}[\alpha]) \leq \lambda(\alpha)$ then there exists a partitionable (i.e. weak* measurable) $f : S \to X$ such that

$$\nu = \int_S f \cdot d\lambda$$

i.e., for every continuous function $\xi : T \to \mathcal{R}$ with compact support,

$$\int_T \xi d\nu = \int_S (\int_T \xi df_s) d\lambda(s) .$$

Proof: For $\alpha \in H$, let $\mu_\alpha \in X$ be the restriction of ν to $\pi^{-1}[\alpha]$, i.e., for any $A \subset T$,

$$\mu_\alpha(A) = \nu(A \cap \pi^{-1}[\alpha]) .$$

Then the $\mu_\alpha/\lambda(\alpha)$ are bounded so 5.1 applies. ∎

Remarks. In the above generality, little can be said about other properties of f. However, in special cases, our methods of proof yield more information.

(1) When S is a separable metric space, π is continuous and λ is a Borel outer measure then we can choose the above f so that, for every $s \in S$, $f(s)$ has its support on $\pi^{-1}[\{s\}]$.

In this case we can avoid the lifting theorem altogether in the proof (of theorem 4.2) by using the simple Vitali system $\mathfrak{J}$ described in example 3 in section 3. Then, for any $s \in S$

and neighborhood β of s, there exists $F \in \mathcal{F}(s)$ such that $\alpha \in F \Rightarrow \alpha \subset \beta$, so the support of $\mu_\alpha/\lambda(\alpha)$ is contained in $\pi^{-1}[\beta]$. Passing to the limit, we conclude that any element in $D(s)$ has its support in $\pi^{-1}\{s\}$.

(2) We may replace the condition 'S is a separable, metric space' in (1) above by "S is a regular topological space and there is a strong lifting on S", i.e. a $\rho : H \to H$ satisfying, in addition to hypotheses (i)-(iv) in theorem 3.3, the requirement $\rho(A) \subset A$ for all closed $A \subset S$.

In this case, if $\mathcal{F}$ is the Vitali system given by theorem 3.3 then again, for any $s \in S$ and closed neighborhood β of s, there exists $F \in \mathcal{F}(s)$ such that $\alpha \in F \Rightarrow \alpha \subset \beta$.

We now turn our attention to results involving the Bochner integral.

5.4. Theorem. Let X be a Banach space, $f : S \to X$ be weakly measurable and, for every $\alpha \in H$,

$$\mu(\alpha) = (\text{Pettis})\text{-}\int_\alpha f \cdot d\lambda .$$

If f takes values in a weakly compact subset of X then $\{\mu(\alpha) \ ; \ \alpha \in H\}$ is relatively (norm)-compact.

<u>Proof:</u> Let C be weakly compact with range $f \subset C$ and B^* be the unit ball in the dual of X endowed with the weak* topology, so B^* is compact. We shall show that $\{\mu(\alpha) ; \alpha \in H\}$ is totally bounded in norm. If this were not the case, suppose $\varepsilon > 0$ and $x_n = \mu(\alpha_n)$ with $\alpha_n \in H$ and $\|x_m - x_n\| > \varepsilon$ for $m,n \in \omega$ and $m \neq n$. Considering $\{x_n ; n \in \omega\}$ as a bounded family of continuous functions on B^*, (by theorem 11.1 on p. 185 in H. H. Schaeffer, Topological Vector Spaces), choose $y_n = x_{i_n}$ so that $\lim_n \psi(y_n)$ exists for every $\psi \in B^*$. Let $\psi_n \in B^*$ with $\psi_n(y_{n+1} - y_n) = \|y_{n+1} - y_n\|$ and (applying the above theorem again) let $\bar{\psi} \in B^*$ and $\xi_n = \psi_{k_n}$ be such that $\lim_n \xi_n(z) = \bar{\psi}(z)$ for every $z \in C$. Then

$$|\xi_n(y_{k_n+1} - y_{k_n})| = |\psi_{k_n}(y_{k_n+1} - y_{k_n})|$$

$$= \|y_{k_n+1} - y_{k_n}\| > \varepsilon \quad .$$

On the other hand,

$$|\xi_n(y_{k_n+1} - y_{k_n})| = |(\xi_n - \bar{\psi})(y_{k_n+1} - y_{k_n}) + \bar{\psi}(y_{k_n+1} - y_{k_n})|$$

$$\leq |(\xi_n - \bar{\psi})(y_{k_n+1})| + |(\xi_n - \bar{\psi})(y_{k_n})| + |\bar{\psi}(y_{k_n+1}) - \bar{\psi}(y_{k_n})|$$

$$\leq 2\int_S |(\xi_n - \bar{\psi})\circ f| d\lambda + |\bar{\psi}(y_{k_n+1}) - \bar{\psi}(y_{k_n})| \quad .$$

Now, by Lebesgue's dominated convergence theorem, for large n,

$$2\int_S |(\xi_n - \bar{\psi})\circ f|\, d\lambda < \frac{\varepsilon}{2} .$$

Also, since $\lim_n \bar{\psi}(y_n)$ exists, for large n,

$$|\bar{\psi}(y_{k_n+1}) - \bar{\psi}(y_{k_n})| < \frac{\varepsilon}{2} .$$

Hence, for large n,

$$\varepsilon < |\xi_n(y_{k_n+1} - y_{k_n})| < \frac{\varepsilon}{2} + \frac{\varepsilon}{2}$$

which is impossible. I

5.5. Theorem. Let X be a Banach space, $\mu << \lambda$ and suppose, for every $\beta' \in H^+$ there exists $\beta \in H^+$ such that $\beta \subset \beta'$ and $\{\mu(\alpha)/\lambda(\alpha) ; \alpha \in H^+$ and $\alpha \subset \beta\}$ is relatively weakly compact. Then there exists a Bochner measurable function $f : S \to X$ such that

$$\mu(\alpha) = \int_\alpha f\cdot d\lambda \qquad \text{for} \quad \alpha \in H .$$

If μ is of bounded variation then

$$\mu(\alpha) = (\text{Bochner})\text{-}\int_\alpha f\cdot d\lambda \quad \text{for } \alpha \in H .$$

<u>Proof:</u> Apply theorem 4.2 and some of the ideas in its proof to get a countable, disjoint $J \subset H^+$ which covers λ-almost all S and a weakly measurable $f : S \to X$ such that, for every $\beta \in J$ and $s \in \beta$,
$E_\beta = \{\mu(\alpha)/\lambda(\alpha) \ ; \ \alpha \in H^+ \text{ and } \alpha \subset \beta\}$ is relatively weakly compact, $f(s) \in$ weak closure E_β and

$$\mu(\alpha) = (\text{Pettis})\text{-}\int_\alpha f \cdot d\lambda \qquad \text{for} \quad \alpha \in H .$$

By theorem 5.2, for each $\beta \in J$,
$\{\mu(\alpha) \ ; \ \alpha \in H \text{ and } \alpha \subset \beta\}$ is relatively (norm)-compact and hence E_β is separable. Let

$$E' = \text{closed convex hull of } \bigcup_{\beta \in J} E_\beta .$$

Then E' is separable and, for λ-almost all $s \in S$, $f(s) \in E'$. It follows that f is partitionable (theorem 3.6 Chap. II) hence Bochner measurable (theorem 2.8 Chap. II) and, for every $\beta \in J$, $\int_\beta \|f\| d\lambda < \infty$ so, for any $\alpha \in H$,

$$\mu(\alpha \cap \beta) = (\text{Bochner})\text{-}\int_{\alpha \cap \beta} f \cdot d\lambda$$

and

$$\mu(\alpha) = \sum_{\beta \in J} (\text{Bochner})\text{-}\int_{\alpha \cap \beta} f \cdot d\lambda .$$

Thus, by theorem 6.2(1) in Chap. II,

$$\mu(\alpha) = \int_\alpha f \cdot d\lambda \quad .$$

If μ is of bounded variation then $\int_S \|f\| d\lambda < \infty$ so f is Bochner integrable. I

As a corollary we get the following.

5.6. Theorem. Let X be a Banach space and U be a bounded linear operator on $\mathcal{L}_1(\lambda)$ to X such that $U[\{\xi \in \mathcal{L}_1(\lambda) : \|\xi\|_1 \leq 1\}]$ is relatively weakly compact in X (i.e., U is a weakly compact or completely continuous operator). Then there exists a Bochner integrable function $f : S \to X$ such that, for every $\xi \in \mathcal{L}_1(\lambda)$,

$$U(\xi) = (\text{Bochner})\text{-}\int_S (\xi \cdot f) d\lambda \quad .$$

Proof: For $\alpha \in H$, let $\mu(\alpha) = U(1_\alpha)$. Then μ is of bounded variation, $\mu << \lambda$, and $\{\mu(\alpha)/\lambda(\alpha) ; \alpha \in H^+\}$ is relatively weakly compact hence, by theorem 5.5, there is a Bochner integrable $f : S \to X$ such that

$$U(1_\alpha) = (\text{Bochner})\text{-}\int_\alpha f \cdot d\lambda \qquad \text{for} \quad \alpha \in H \quad .$$

Since the simple functions are dense in $\mathcal{L}_1(\lambda)$, we conclude

$$U(\xi) = (\text{Bochner})\text{-}\int_S (\xi \cdot f)\, d\lambda \quad \text{for} \quad \xi \in \mathcal{L}_1(\lambda) \ . \quad \mathrm{I}$$

Note: Again, as in theorem 5.2, it follows immediately from our construction of f that

$$\sup_{s \in S} \|f(s)\| = \|U\|.$$

REFERENCES

1. Dunford N. and Pettis, J. B. Linear operations on summable functions. Trans. A.M.S. 47 (1940), pp. 323-392.

2. Ionescu Tulcea, A. and C. Topics in the theory of lifting. Springer-Verlag, Heidelberg-New York, 1969.

3. Kölzow, D. Differentiation von Massen. Lecture Notes in Mathematics 65, Springer 1968.

4. Pellaumail, J. Sur la dérivation des mesures vectorielles. C.R. Acad. Sci. Paris 269 (1969), pp. 904-907.

5. Pettis, J. B. On integration in vector spaces. Trans. A.M.S. 44 (1938), pp. 277-304.

6. ______________, Differentiation in Banach spaces. Duke Math. J. 5 (1939), pp. 254-269.

7. de Possel, R. Sur la dérivation abstraite des fonctions d'ensembles. J. de Math. Pures et Appl. 15 (1936), pp. 391-409.

8. Phillips, R. S. On weakly compact subsets of a Banach space. Amer. J. Math. 65 (1943), pp. 108-136.

9. Sion, M. Lectures on vector-valued measures. University of British Columbia, 1969-70.

10. ______________, A proof of the lifting theorem. University of British Columbia, 1970.

11. ______________, Group-valued outer measures. Actes du Congrès International des Mathématiciens, Nice 1970, vol. 2, pp. 589-593.

12. Traynor, T. Differentiation of group-valued outer measures. Thesis, University of British Columbia, 1969.

The notion of completely discontinuous or singular measure, in the vector-valued case, is due to Traynor [12].

The notion of a differentiation basis in an abstract space is due to de Possel [7]. There is a vast literature on the subject. The fact that the lifting theorem yields a suitable differentiation basis was noted independently by Kölzow [3], by Sion and Traynor [10, 11, 12], and by Pellaumail [4]. However, the lifting theorem had been used extensively before by A. and C. Ionescu Tulcea in connection with integral representation theorems. Most of their results are collected in [2]. Their methods however are essentially functional analytic in character.

Theorem 5.1, with an added separability condition, is due to Pettis [5, 6] while its corollary 5.2 (also with added restrictions) is due to Dunford and Pettis [1].

Theorem 5.4 does not seem to have been formulated explicitely in the literature before. However, it can be found in Phillips ([8], p. 131) as part of the proof of a slightly modified version of theorem 5.5. It should be pointed out that our proof of theorem 5.4 (and hence 5.5) uses only a version of Eberlein's theorem for real-valued functions and the weak* compactness of the unit ball (Alaoglu's theorem). Incidentally, the proof of theorem 5.5 in Phillips seems to have a gap since the family of partitions by sets of positive measure is not directed by refinement. This gap is now easily plugged by using the lifting theorem as done by Pellaumail in [4].

There is an extensive literature on disintegration of measures (see e.g. [2]). Our version (theorem 5.3) does not seem to have been formulated before. Our method of proof, as a corollary to 5.1, also seems new.

A general differentiation theory when the base measure λ is also vector-valued has been considered by Traynor [12].

APPENDIX

THE LIFTING THEOREM

In this appendix we give a simple direct proof of the lifting theorem, in the case of a finite measure, which clearly brings out its connection with differentiation. The methods we follow are close in spirit to those originally used by von Neumann working with Lebesgue measure on the real line. For a good guide to the literature on this subject, we refer the reader to the book by A. and C. Ionescu Tulcea, 'Topics in the Theory of Lifting', Springer-Verlag, 1969.

We assume only acquaintance with the basic notions of Caratheodory outer measures on an abstract space, thereby making this appendix essentially self contained.

0. Basic Notation

Throughout this appendix,

S is an abstract space;

λ is a non-negative outer measure on S with $\lambda(S) < \infty$;

$\mathcal{M}$ is the family of λ-measurable sets;

$\mathcal{M}^+ = \{A \in \mathcal{M}: \ 0 < \lambda(A) < \lambda(S)\} \cup \{\phi, S\}$;

$\mathcal{N} = \{A: \lambda(A) = 0\}$;

$\omega = \{0, 1, 2, \ldots\}$.

For any subsets A, B of S,

$A \setminus B = \{x \in A: \ x \notin B\}$;

$\sim B = S \setminus B$;

$A \equiv B$ iff $\lambda((A \setminus B) \cup (B \setminus A)) = 0$.

For any family F of subsets of S,

$$\bigcup F = \bigcup_{\alpha \in F} \alpha,$$

Borelfield F = smallest σ-field containing F;

B_F = Borelfield $(F \cup \mathcal{N})$

(= completion of Borelfield F with respect to λ);

λ_F = The outer measure generated by $\lambda / (F \cup \mathcal{N})$, i.e. for any $A \subset S$,

$$\lambda_F(A) = \inf \{ \sum_{\alpha \in H} \lambda(\alpha); \ \ H \text{ is countable, } H \subset F \ , A \subset \bigcup H\};$$

F is complete iff, for every $A \subset S$ and $B \in F$, if $A \equiv B$ then $A \in F$.

Liftings

Definitions

For any field $\mathcal{B} \subset \mathfrak{M}$,

(1) ρ is a density function on $\mathcal{B}$ iff $\rho : \mathcal{B} \to \mathfrak{M}^+$

and, for every $A, B \in \mathcal{B}$,

(i) $A \equiv \rho(A)$

(ii) $A \equiv B \Rightarrow \rho(A) = \rho(B)$

(iii) $\rho(A \cap B) = \rho(A) \cap \rho(B)$

Note that (i) and the definition of $\mathfrak{M}^+$ imply

$\rho(\phi) = \phi$ and $\rho(S) = S$.

(2) ρ is a lifting on $\mathcal{B}$ iff ρ is a density function on $\mathcal{B}$ which satisfies the added condition

(iv) $\rho(\sim A) = \sim\rho(A)$ for $A \in \mathcal{B}$.

(3) F lifts $\mathcal{B}$ iff $F \subset \mathfrak{M}^+$, F is a field and, for every $A \in \mathcal{B}$, there is a (necessarily unique) $\alpha \in F$ with $A \equiv$

(4) F is a partial lifting iff F lifts $\mathcal{B}_F$.

Note: If ρ is a lifting on $\mathcal{B}$ then range ρ lifts $\mathcal{B}$. Conversely, if F lifts $\mathcal{B}$ and, for every $A \in \mathcal{B}$, $\rho(A)$ is the $\alpha \in F$ with $A \equiv \alpha$ then ρ is a lifting on $\mathcal{B}$.

2. Differentiation Systems

Definitions

(1) F is a <u>differentiation system</u> iff $F \subset \mathfrak{M}^+$ and, for every $\alpha, \beta \in F$, we have $\alpha \cap \beta \in F$.

(2) For any differentiation system F and $s \in S$,

$\hat{F}(s) = \{\alpha \in F: \; s \in \alpha\}$

and $\hat{F}(s)$ is directed by inclusion downward.

(3) For any differentiation system F, $H \subset F$ and $A \subset S$, H is a Vitali cover for A iff, for every $s \in A$ and $\beta \in \hat{F}(s)$, there exists $\alpha \in H$ with $s \in \alpha \subset \beta$.

(4) F is a <u>Vitali system</u> iff F is a differentiation system such that

(i) for every $A \subset S$ and Vitali cover H for A, there exists a countable, disjoint $H' \subset H$ with $\lambda(A \setminus \bigcup H') = 0$;

(ii) $\lambda_F \,/\, \mathcal{B}_F = \lambda \,/\, \mathcal{B}_F$.

3. Main Steps to the Lifting Theorem

Here we only list the main steps leading to the lifting theorem. Their proofs are given in the next section.

Theorem 1 If F is a partial lifting then F is a Vitali system.

Theorem 2 If F is a Vitali system, $T \subset S$, $r \geq 0$,

$$D_* = \{s \in S : \underline{\lim}_{\alpha \in \hat{F}(s)} \frac{\lambda(\alpha \cap T)}{\lambda(\alpha)} \leq r\},$$

$$D^* = \{s \in S : \overline{\lim}_{\alpha \in \hat{F}(s)} \frac{\lambda(\alpha \cap T)}{\lambda(\alpha)} \geq r\},$$

then $D_* \in \mathcal{B}_F$, $D^* \in \mathcal{B}_F$ and, for any $B \in \mathcal{B}_F$,

$$\lambda(B \cap D_* \cap T) \leq r \cdot \lambda(B \cap D_*)$$

$$\lambda(B \cap D^* \cap T) \geq r \cdot \lambda(B \cap D^*).$$

Theorem 3 If, for each $n \in \omega$, F_n is a Vitali system, $F_n \subset F_{n+1}$, $A \in \text{Borelfield}\,(\bigcup_{n\in\omega} \mathcal{B}_{F_n})$, and

$$B = \{s \in S : \lim_n \underline{\lim}_{\alpha \in \hat{F}_n(s)} \frac{\lambda(\alpha \cap A)}{\lambda(\alpha)} = 1\}$$

then $A \equiv B$.

Theorem 4 If, for each $n \in \omega$, F_n is a partial lifting and $F_n \subset F_{n+1}$ then there exists a density function ρ on Borelfield $(\bigcup_{n\in\omega} \mathcal{B}_{F_n})$ with $\rho(A) = A$ for $A \in \bigcup_{n\in\omega} F_n$.

Theorem 5 If ρ is a density function on a complete field $\mathcal{B} \subset \mathfrak{M}$ then there exists an F which lifts $\mathcal{B}$ and such that

$$\rho(A) \subset A \subset \sim\rho(\sim A) \qquad \text{for } A \in F.$$

Corollary If, for each $n \in \omega$, F_n is a partial lifting and $F_n \subset F_{n+1}$ then there exists a partial lifting H with $\bigcup_{n\in\omega} F_n \subset H$.

Lemma If F is a partial lifting and $A \in \mathfrak{M} \setminus \mathcal{B}_F$ then there exists a partial lifting F' such that $F \subset F'$ and $A \in \mathcal{B}_{F'}$.

<u>Lifting Theorem</u> There exists a maximal partial lifting F with $\mathcal{B}_F = \mathfrak{M}$. Thus, there exists a lifting on $\mathfrak{M}$.

4. Proofs

Here we give the proofs of the theorems listed in the previous section. In fact, we prove somewhat stronger results, for the sake of perspective.

Theorem 1 If F is a partial lifting then F is a Vitali system.

Proof: (i) Given $A \subset S$ and a Vitali cover H for A, let H' be a maximal, disjoint subfamily of H. Then H' is countable and there exists $\beta \in F$ with $(S \setminus \bigcup H') \equiv \beta$. For every $\alpha \in H'$, we have $\alpha \cap \beta \equiv \phi$ and $\alpha \cap \beta \in F \subset \mathfrak{M}^+$ hence $\alpha \cap \beta = \phi$. Therefore, we cannot have any $s \in A \cap \beta$ for, otherwise, there would be an $\alpha' \in H$ with $s \in \alpha' \subset \beta$ contradicting the maximality of H'. Thus, $A \cap \beta = \phi$ and so $\lambda(A \setminus \bigcup H') = 0$.

(ii) Since F lifts $\mathcal{B}_F$ clearly $\lambda_F \,/\, \mathcal{B}_F = \lambda \,/\, \mathcal{B}_F$.

Theorem 2' Let F be a Vitali system, $r \geq 0$, ν be a finite measure on $\mathcal{B}_F$ which is absolutely continuous with respect to $\lambda/\mathcal{B}_F$,

$$D_* = \{s \in S : \varliminf_{\alpha \in \hat{F}(s)} \frac{\nu(\alpha)}{\lambda(\alpha)} \leq r\},$$

$$D^* = \{s \in S : \varlimsup_{\alpha \in \hat{F}(s)} \frac{\nu(\alpha)}{\lambda(\alpha)} \geq r\}.$$

If $\bar{\nu}$ is the outer measure generated by ν then

(1) $A \subset D_* \Rightarrow \bar{\nu}(A) \leq r \cdot \lambda_F(A)$

(2) $A \subset D^* \Rightarrow \bar{\nu}(A) \geq r \cdot \lambda_F(A)$

(3) $D_* \in \mathcal{B}_F$ and $D^* \in \mathcal{B}_F$.

Proof: For any $r' > r$ and $B \in F_\sigma$, let

$$H = \{\alpha \in F : \alpha \subset B \text{ and } \frac{\nu(\alpha)}{\lambda(\alpha)} < r'\}.$$

Then H is a Vitali cover for $B \cap D_*$, hence there exists a countable, disjoint $H' \subset H$ with $\lambda(B \cap D_* \setminus \bigcup H') = 0$. So $\nu(B \cap D_* \setminus \bigcup H') = 0$ and $\bar{\nu}(B \cap D_* \setminus \bigcup H') = 0$ and

$$\bar{\nu}(B \cap D_*) \leq \sum_{\alpha \in H'} \nu(\alpha) \leq r' \cdot \sum_{\alpha \in H'} \lambda(\alpha) \leq r' \cdot \lambda(B).$$

Given any $A \subset S$ and $\varepsilon>0$, from the definitions of λ_F and $\bar{\nu}$ and absolute continuity of ν, there exists $B \in F_\sigma$ such that

$\lambda(A \setminus B) = 0$ so $\nu(A \setminus B) = 0$ and $\bar{\nu}(A \setminus B) = 0$, and

$\lambda(B) \leq \lambda_F(A) + \varepsilon.$

If $A \subset D_*$ then

$$\bar{\nu}(A) \leq \bar{\nu}(B \cap D_*) \leq r' \cdot \lambda(B) \leq r' \cdot \lambda_F(A) + r' \cdot \varepsilon.$$

Letting $r' \to r$ and $\varepsilon \to 0$, we get

(1) $A \subset D_* \Rightarrow \bar{\nu}(A) \leq r \cdot \lambda_F(A)$

Similarly, we get

(2) $A \subset D^* \Rightarrow \bar{\nu}(A) \geq r \cdot \lambda_F(A).$

To check (3), let B be a common λ_F, $\bar{\nu}$ - outer hull of D_*, i.e., let $B \in \mathcal{B}_F$, $D_* \subset B$, $\lambda_F(D_*) = \lambda(B)$ and $\bar{\nu}(D_*) = \nu(B)$. Then, for every $\alpha \in \mathcal{B}_F$ with $\alpha \subset B$, we have

$$\lambda(\alpha) = \lambda_F(\alpha \cap D_*) \text{ and } \nu(\alpha) = \bar{\nu}(\alpha \cap D_*).$$

To see that $\lambda(B \setminus D_*) = 0$, let

$$A_n = \{s \in B \setminus D_* : \overline{\lim_{\alpha \in \hat{F}(s)}} \frac{\nu(\alpha)}{\lambda(\alpha)} > r + \frac{1}{n}\}$$

and choose $\alpha_n \in \mathcal{B}_F$ so that $A_n \subset \alpha_n \subset B$, $\lambda(\alpha_n) = \lambda_F(A_n)$ and $\nu(\alpha_n) = \bar{\nu}(A_n)$. Then, by (2),

$$\bar{\nu}(A_n) \geq (r + \frac{1}{n}) \cdot \lambda_F(A_n)$$

On the other hand, with the help of (1), we have

$$\bar{\nu}(A_n) = \nu(\alpha_n) = \bar{\nu}(\alpha_n \cap D_*) \leq r \cdot \lambda_F(\alpha_n \cap D_*) = r \cdot \lambda(\alpha_n)$$
$$= r \cdot \lambda_F(A_n).$$

Therefore, $\lambda_F(A_n) = 0$ and, since $B \setminus D_* \subset \bigcup_{n \in \omega} A_n$, we conclude

$\lambda(B \setminus D_*) = 0$ so $D_* \varepsilon \mathcal{B}_F$. Similarly, $D^* \varepsilon \mathcal{B}_F$.

Theorem 2 is then an immediate corollary of theorem 2' obtained by letting $\nu(A) = \lambda(A \cap T)$ for $A \varepsilon \mathcal{B}_F$.

Theorem 3 If, for each $n\varepsilon\omega$, F_n is a Vitali system, $F_n \subset F_{n+1}$, $\mathcal{B}$ = Borelfield $(\bigcup_{n\varepsilon\omega} \mathcal{B}_{F_n})$, $A \varepsilon \mathcal{B}$, and

$$B = \{s \varepsilon S : \lim_n \underline{\lim}_{\alpha\varepsilon\hat{F}_n(s)} \frac{\lambda(\alpha \cap A)}{\lambda(\alpha)} = 1\}$$

then $A \equiv B$.

Proof: For any $r < 1$, let

$$D_*(n) = \{s \varepsilon S : \underline{\lim}_{\alpha\varepsilon\hat{F}_n(s)} \frac{\lambda(\alpha \cap A)}{\lambda(\alpha)} < r\},$$

$$E_* = E_*(r) = \bigcap_{N\varepsilon\omega} \bigcup_{n\geq N} D_*(n).$$

Then $(A \setminus B) \subset \bigcup_{r<1}(A \cap E_*(r))$.

To see that $\lambda(A \cap E_*) = 0$, let $N\varepsilon\omega$, $\beta \varepsilon \mathcal{B}_{F_N}$ and, for $n \geq N$,

$$D'_n = D_*(n) \setminus \bigcup_{j=N}^{n-1} D_*(j).$$

Then $\beta \cap D'_n \varepsilon \mathcal{B}_{F_n}$ and so by theorem 2,

$$\lambda(\beta \cap D'_n \cap A) \leq r \cdot \lambda(\beta \cap D'_n).$$

Summing over $n \geq N$, we get

$$\lambda(\beta \cap \bigcup_{n\geq N} D_*(n) \cap A) \leq r \cdot \lambda(\beta \cap \bigcup_{n\geq N} D_*(n)),$$

hence

$$\lambda(\beta \cap E_* \cap A) \leq r \cdot \lambda(\beta \cap E_*).$$

By considering monotone sequences of such β's, we conclude that the above inequality holds for any $\beta \in \mathcal{B}$. In particular then, letting $\beta = A$, we get $\lambda(A \cap E_*) \leq r \cdot \lambda(A \cap E_*)$ and, since $r < 1$, we must have $\lambda(A \cap E_*) = 0$. Letting $r \to 1$ through an increasing sequence, we conclude $\lambda(A \setminus B) = 0$. Similarly, for any $r > 0$, if

$$D^*(n) = \{s \in S : \overline{\lim_{\alpha \in \hat{F}_n(s)}} \frac{\lambda(\alpha \cap A)}{\lambda(\alpha)} > r\},$$

$$E^* = \bigcap_{N \in \omega} \bigcup_{n \geq N} D^*(n),$$

and $\beta \in \mathcal{B}$ then

$$\lambda(\beta \cap E^* \cap A) \geq r \cdot \lambda(\beta \cap E^*).$$

Letting $\beta = \sim A$, we get $\lambda(E^* \setminus A) = 0$ for every $r > 0$, so for λ - a.a. $s \in (S \setminus A)$,

$$\lim_n \overline{\lim_{\alpha \in \hat{F}_n(s)}} \frac{\lambda(\alpha \cap A)}{\lambda(\alpha)} = 0.$$

Thus, $\lambda(B \setminus A) = 0$ and $A \equiv B$.

Theorem 4 If, for each $n \in \omega$, F_n is a partial lifting, $F_n \subset F_{n+1}$ and $\mathcal{B} = \text{Borelfield}\ (\bigcup_{n \in \omega} \mathcal{B}_{F_n})$ then there exists a density function ρ on $\mathcal{B}$ with $\rho(A) = A$ for $A \in \bigcup_{n \in \omega} F_n$.

Proof: For every $A \in \mathcal{B}$, let

$$\rho(A) = \{s \in S : \lim_n \underline{\lim_{\alpha \in F_n(s)}} \frac{\lambda(\alpha \cap A)}{\lambda(\alpha)} = 1\}.$$

By theorem 3, $A \equiv \rho(A)$ and, since each F_n is a field,

$$A \in \bigcup_{n \in \omega} F_n \Rightarrow \rho(A) = A.$$

For any $A, B \in \mathcal{B}$, we trivially have

$$A \equiv B \Rightarrow \rho(A) = \rho(B).$$

Finally, to see that $\rho(A \cap B) = \rho(A) \cap \rho(B)$,

let

$$f_n^A(s) = \underset{\alpha\varepsilon\hat{F}_n(s)}{\underline{\lim}} \frac{\lambda(\alpha \cap A)}{\lambda(\alpha)}$$

$$\overline{f}_n^A(s) = \underset{\alpha\varepsilon\hat{F}_n(s)}{\overline{\lim}} \frac{\lambda(\alpha \cap A)}{\lambda(\alpha)}$$

$$g^A(s) = \underset{n}{\underline{\lim}}\ f_n^A(s)$$

$$\overline{g}^A(s) = \underset{n}{\overline{\lim}}\ \overline{f}_n^A(s)$$

and note:

$$s \ \varepsilon\ \rho(A) \Longleftrightarrow g^A(s) = 1.$$

(i) $g^{A \cap B}(s) = 1 \Rightarrow\ 1 = g^{A \cap B}(s) \leq g^A(s) \leq 1$ and

$$1 = g^{A \cap B}(s) \leq g^B(s) \leq 1$$

$$\Rightarrow\ g^A(s) = 1 \ \text{ and } \ g^B(s) = 1$$

so $\qquad \rho(A \cap B) \subset \rho(A) \cap \rho(B).$

On the other hand, using the elementary inequality

$$\underline{\lim}\ (a + b) \leq \underline{\lim}\ a + \overline{\lim}\ b \leq \overline{\lim}\ (a + b)$$

we have:

(ii) $f_n^A(s) + \overline{f}_n^{(B \setminus A)}(s) \leq \overline{f}_n^{(A \cup B)}(s)$

hence

$$g^A(s) = 1 \Rightarrow \overline{g}^{(A \cup B)}(s) = 1$$

$$\Rightarrow 0 \leq \overline{g}^{(B \setminus A)}(s) \leq \overline{g}^{(A \cup B)}(s) - g^A(s) = 0;$$

(iii) $f_n^B(s) \leq f_n^{(A \cap B)}(s) + \overline{f}_n^{(B \setminus A)}(s)$

hence, using (ii),

$$g^A(s) = 1 \text{ and } g^B(s) = 1 \Rightarrow g^{(A \cap B)}(s) \geq g^B(s) - \overline{g}^{(B \setminus A)}(s) =$$

$$\Rightarrow g^{(A \cap B)}(s) = 1$$

so $\rho(A) \cap \rho(B) \subset \rho(A \cap B)$.

<u>Theorem 5</u> If ρ is a density function on a complete field $\mathcal{B} \subset \mathcal{M}$ then there exists an F which lifts $\mathcal{B}$ and such that

$$\rho(A) \subset A \subset \sim\rho(\sim A) \quad \text{for } A \in F.$$

<u>Proof</u>: Let $\rho^*(A) = \sim\rho(\sim A)$. Then, for any $A, B \in \mathcal{B}$,

(i) $A \equiv \rho^*(A)$

(ii) $\rho^*(A \cup B) = \rho^*(A) \cup \rho^*(B)$

since $\sim A \equiv \rho(\sim A)$ and $\rho(\sim A \cap \sim B) = \rho(\sim A) \cap \rho(\sim B)$.

Let

$$\Delta = \{F : F \text{ is a field, } F \subset \mathcal{B} \text{ and, for every } \alpha \in F,$$
$$\rho(\alpha) \subset \alpha \subset \rho^*(\alpha)\}$$

and F be a maximal element in Δ. To check that F lifts $\mathcal{B}$, given $A \in \mathcal{B}$, let

$$A' = \bigcup_{\alpha \in F} [\rho(\alpha \cup A) \setminus \alpha].$$

We shall show that $A \equiv A'$ and $A' \in F$. First note:

(1) $\alpha, \beta \in F \Rightarrow [\rho(\alpha \cup A) \setminus \alpha] \cap [\rho(\beta \cup \sim A) \setminus \beta] = \phi$

for, $\rho(\alpha \cup A) \cap \rho(\beta \cup \sim A) \cap \sim\alpha \cap \sim\beta \subset \rho(\alpha \cup \beta) \cap \sim(\alpha \cup \beta) = \phi$.

(2) $\beta \in F \Rightarrow A' \cap [\rho(\beta \cup \sim A) \setminus \beta] = \phi$

(3) $\rho(A) \subset A' \subset \rho^*(A)$ so by (i) $A \equiv A'$

for, $\rho(A) \subset A'$ from definition of A' since $\phi \in F$,

$A' \subset \rho^*(A)$ from (2) with $\beta = \phi$.

(4) $\beta \in F \Rightarrow \beta \cap A' \subset \rho^*(\beta \cap A')$

for, from (3) and (2), $\rho(\sim\beta \cup \sim A') = \rho(\sim\beta \cup \sim A) \subset \sim\beta \cup \sim A'$.

(5) $\gamma \in F \Rightarrow \gamma \setminus A' \subset \rho^*(\gamma \setminus A')$

for, $\rho(\sim\gamma \cup A') = \rho(\sim\gamma \cup A) \subset \sim\gamma \cup A'$.

Thus, if F' is the field generated by $F \cup \{A'\}$ so that

$$F' = \{\alpha \cup (\beta \cap A') \cup (\gamma \setminus A');\ \alpha, \beta, \gamma \in F\}$$

then, by (ii), (4) and (5), for any $C \in F'$, we have

$C \subset \rho^*(C)$ and $\sim C \subset \rho^*(\sim C)$ hence $\rho(C) \subset C \subset \rho^*(C)$.

Thus, $F \subset F' \in \Delta$ and therefore $F = F'$ and $A' \in F$.

Lemma If F is a partial lifting and $A \in (\mathcal{M} \setminus \mathcal{B}_F)$, then there exists a partial lifting F' with $F \subset F'$ and $A \in \mathcal{B}_{F'}$.

Proof: Let

$$F_1 = \{\alpha \in F : (\alpha \cap A) \equiv \phi\}$$
$$F_2 = \{\alpha \in F : (\alpha \setminus A) \equiv \phi\}$$

then choose $B_1 \in (F_1)_\sigma$ and $B_2 \in (F_2)_\sigma$ with

$$\lambda(B_1) = \sup \{\lambda(\alpha) \; ; \; \alpha \in F_1\}$$
$$\lambda(B_2) = \sup \{\lambda(\alpha) \; ; \; \alpha \in F_2\}$$

and

$$\beta_1, \beta_2 \in F \text{ with } B_1 \equiv \beta_1 \text{ and } B_2 \equiv \beta_2.$$

We must have $\beta_1 \in F_1$, $\beta_2 \in F_2$,

$\bigcup F_1 = \beta_1$ since $\alpha \in F_1 \Rightarrow \lambda(\alpha \setminus \beta_1) = 0 \Rightarrow (\alpha \setminus \beta_1) = \phi$

$\bigcup F_2 = \beta_2$ since $\alpha \in F_2 \Rightarrow \lambda(\alpha \setminus \beta_2) = 0 \Rightarrow (\alpha \setminus \beta_2) = \phi$

$\beta_1 \cap \beta_2 = \phi$ since $(\beta_1 \cap \beta_2) \subset (\beta_1 \cap A) \cup (\beta_2 \setminus A) \equiv \phi$

Let $A' = (A \setminus \beta_1) \cup \beta_2$ so $A \equiv A'$ and

$F' = \{\alpha \cup (\beta \cap A') \cup (\gamma \setminus A'); \alpha, \beta, \gamma \in F\}$.

Then F' is a field, $F \subset F'$ and $F' \subset \mathcal{M}$ since, for $\beta, \gamma \in F$:

$(\beta \cap A') \equiv \phi \Rightarrow \beta \in F_1 \Rightarrow \beta \subset \beta_1 \Rightarrow \beta \cap A' = \phi$

$(\gamma \setminus A') \equiv \phi \Rightarrow \gamma \in F_2 \Rightarrow \gamma \subset \beta_2 \Rightarrow \gamma \setminus A' = \phi$.

Since $\mathcal{B}_{F'} = \{\alpha \cup (\beta \cap A) \cup (\gamma \setminus A); \alpha, \beta, \gamma \in \mathcal{B}_F$

we see that F' lifts $\mathcal{B}_{F'}$ and $A \in \mathcal{B}_{F'}$.

Lifting Theorem There exists a lifting on $\mathcal{M}$.

Proof: Let $\mathcal{N}$ be a maximal nest of partial liftings and $L = \bigcup\mathcal{N}$. Then L is a partial lifting, for

(1) If $\mathcal{N}$ has no cofinal sequence then

$$\mathcal{B}_L = \bigcup_{F \in \mathcal{N}} \mathcal{B}_F$$

so L lifts $\mathcal{B}_L$.

(2) If $\mathcal{N}$ has a cofinal sequence then, by the corollary to theorems 4 and 5, there is a partial lifting F' with $L \subset F'$ and hence $L = F'$.

Since L is a maximal partial lifting, by the lemma we must have $\mathcal{M} = \mathcal{B}_L$. Thus, L lifts $\mathcal{M}$. If, for $A \in \mathcal{M}$, $\rho(A)$ is the $\alpha \in L$ with $A \equiv \alpha$ then ρ is a lifting on $\mathcal{M}$.

Remarks

(1) Theorem 1 above is a special case of theorem 3.3 in Chapter III. The historical remarks at the end of Chapter III (p.123) apply here too.

(2) Theorem 2 is a well known classical result whereas theorem 3 is a simple version of the Martingale theorem.

(3) The material in this appendix is taken from my Lecture Notes "A Proof of the Lifting Theorem", University of British Columbia, March 1970. I am indebted to J. Kupka for pointing out minor errors in these notes and suggesting a way to avoid the use of universal subnets in the proof of theorem 4. The proof given here is partially based on his suggestions.

5. Lifting and Differentiation

The essence of the relation between the notions of lifting and differentiation is contained in theorem 1. The nature of the relation becomes clearer if one realizes that theorem 2' is a key step in the Lebesgue approach to differentiation, for one readily concludes from it that, if

$$f(s) = \lim_{\alpha \varepsilon \hat{F}(s)} \frac{\nu(\alpha)}{\lambda(\alpha)} ,$$

then f(s) exists for λ-almost all s ε S, f is a $\mathcal{B}_F$-measurable function and, for every A ε $\mathcal{B}_F$,

$$\nu(A) = \int_A f \, d\lambda.$$

With the help of this, theorem 1 can be sharpened to yield:

Theorem 1'

(i) If ρ is a density function on a complete σ-field $\mathcal{B}$ and F = range ρ, then F is a Vitali system with $\mathcal{B}_F = \mathcal{B}$.

(ii) If F is a Vitali system then there exists a density function ρ on $\mathcal{B}_F$ with $A \subset \rho(A)$ for every A ε F.

Proof:

(i) Same as theorem 1, since the fact that A ε F => ∼A ε F was not used there.

(ii) Let

$$\rho(A) = \{s \varepsilon S : \lim_{\alpha \varepsilon \hat{F}(s)} \frac{\lambda(\alpha \cap A)}{\lambda(\alpha)} = 1\}$$

(Note this is a very special case of theorem 4).

With the help of theorem 5, we then have

<u>Theorem 1"</u>

(i) If F lifts a complete field $\mathcal{B}$ then F is a Vitali system with $\mathcal{B}_F = \mathcal{B}$.

(ii) If F is a Vitali system then there exists an F' which lifts $\mathcal{B}_F$.

The lifting theorem then can be restated in the form

<u>Theorem</u> There exists a Vitali system F with $\mathcal{M} = \mathcal{B}_F$.

This points out the significance of the lifting theorem for differentiation: it provides us with a Vitali system in a general measure space, to play the role of the family of intervals on the line, and thereby enables us to follow the classical Lebesgue approach in differentiating a measure ν with respect to λ to obtain an integral representation for ν. What makes this approach very useful is that, unlike the situation with the Radon-Nikodym theorem, the Vitali system produced here depends only on the base measure λ and <u>not</u> on the measure ν being differentiated. As a result, by this approach one can obtain integral representations for a vector-valued measure ν even when ν has, in an essential way, an uncountable number of coordinates.

Vol. 215: P. Antonelli, D. Burghelea and P. J. Kahn, The Concordance-Homotopy Groups of Geometric Automorphism Groups. X, 140 pages. 1971. DM 16,–

Vol. 216: H. Maaß, Siegel's Modular Forms and Dirichlet Series. VII, 328 pages. 1971. DM 20,–

Vol. 217: T. J. Jech, Lectures in Set Theory with Particular Emphasis on the Method of Forcing. V, 137 pages. 1971. DM 16,–

Vol. 218: C. P. Schnorr, Zufälligkeit und Wahrscheinlichkeit. IV, 212 Seiten. 1971. DM 20,–

Vol. 219: N. L. Alling and N. Greenleaf, Foundations of the Theory of Klein Surfaces. IX, 117 pages. 1971. DM 16,–

Vol. 220: W. A. Coppel, Disconjugacy. V, 148 pages. 1971. DM 16,–

Vol. 221: P. Gabriel und F. Ulmer, Lokal präsentierbare Kategorien. V, 200 Seiten. 1971. DM 18,–

Vol. 222: C. Meghea, Compactification des Espaces Harmoniques. III, 108 pages. 1971. DM 16,–

Vol. 223: U. Felgner, Models of ZF-Set Theory. VI, 173 pages. 1971. DM 16,–

Vol. 224: Revètements Etales et Groupe Fondamental. (SGA 1). Dirigé par A. Grothendieck XXII, 447 pages. 1971. DM 30,–

Vol. 225: Théorie des Intersections et Théorème de Riemann-Roch. (SGA 6). Dirigé par P. Berthelot, A. Grothendieck et L. Illusie. XII, 700 pages. 1971. DM 40,–

Vol. 226: Seminar on Potential Theory, II. Edited by H. Bauer. IV, 170 pages. 1971. DM 18,–

Vol. 227: H. L. Montgomery, Topics in Multiplicative Number Theory. IX, 178 pages. 1971. DM 18,–

Vol. 228: Conference on Applications of Numerical Analysis. Edited by J. Ll. Morris. X, 358 pages. 1971. DM 26,–

Vol. 229: J. Väisälä, Lectures on n-Dimensional Quasiconformal Mappings. XIV, 144 pages. 1971. DM 16,–

Vol. 230: L. Waelbroeck, Topological Vector Spaces and Algebras. VII, 158 pages. 1971. DM 16,–

Vol. 231: H. Reiter, L^1-Algebras and Segal Algebras. XI, 113 pages. 1971. DM 16,–

Vol. 232: T. H. Ganelius, Tauberian Remainder Theorems. VI, 75 pages. 1971. DM 16,–

Vol. 233: C. P. Tsokos and W. J. Padgett. Random Integral Equations with Applications to stochastic Systems. VII, 174 pages. 1971. DM 18,–

Vol. 234: A. Andreotti and W. Stoll. Analytic and Algebraic Dependence of Meromorphic Functions. III, 390 pages. 1971. DM 26,–

Vol. 235: Global Differentiable Dynamics. Edited by O. Hájek, A. J. Lohwater, and R. McCann. X, 140 pages. 1971. DM 16,–

Vol. 236: M. Barr, P. A. Grillet, and D. H. van Osdol. Exact Categories and Categories of Sheaves. VII, 239 pages. 1971. DM 20,–

Vol. 237: B. Stenström, Rings and Modules of Quotients. VII, 136 pages. 1971. DM 16,–

Vol. 238: Der kanonische Modul eines Cohen-Macaulay-Rings. Herausgegeben von Jürgen Herzog und Ernst Kunz. VI, 103 Seiten. 1971. DM 16,–

Vol. 239: L. Illusie, Complexe Cotangent et Déformations I. XV, 355 pages. 1971. DM 26,–

Vol. 240: A. Kerber, Representations of Permutation Groups I. VII, 192 pages. 1971. DM 18,–

Vol. 241: S. Kaneyuki, Homogeneous Bounded Domains and Siegel Domains. V, 89 pages. 1971. DM 16,–

Vol. 242: R. R. Coifman et G. Weiss, Analyse Harmonique Non-Commutative sur Certains Espaces. V, 160 pages. 1971. DM 16,–

Vol. 243: Japan-United States Seminar on Ordinary Differential and Functional Equations. Edited by M. Urabe. VIII, 332 pages. 1971. DM 26,–

Vol. 244: Séminaire Bourbaki – vol. 1970/71. Exposés 382–399. IV, 356 pages. 1971. DM 26,–

Vol. 245: D. E. Cohen, Groups of Cohomological Dimension One. V, 99 pages. 1972. DM 16,–

Vol. 246: Lectures on Rings and Modules. Tulane University Ring and Operator Theory Year, 1970–1971. Volume I. X, 661 pages. 1972. DM 40,–

Vol. 247: Lectures on Operator Algebras. Tulane University Ring and Operator Theory Year, 1970–1971. Volume II. XI, 786 pages. 1972. DM 40,–

Vol. 248: Lectures on the Applications of Sheaves to Ring Theory. Tulane University Ring and Operator Theory Year, 1970–1971. Volume III. VIII, 315 pages. 1971. DM 26,–

Vol. 249: Symposium on Algebraic Topology. Edited by P. J. Hilton. VII, 111 pages. 1971. DM 16,–

Vol. 250: B. Jónsson, Topics in Universal Algebra. VI, 220 pages. 1972. DM 20,–

Vol. 251: The Theory of Arithmetic Functions. Edited by A. A. Gioia and D. L. Goldsmith VI, 287 pages. 1972. DM 24,–

Vol. 252: D. A. Stone, Stratified Polyhedra. IX, 193 pages. 1972. DM 18,–

Vol. 253: V. Komkov, Optimal Control Theory for the Damping of Vibrations of Simple Elastic Systems. V, 240 pages. 1972. DM 20,–

Vol. 254: C. U. Jensen, Les Foncteurs Dérivés de $\varprojlim$ et leurs Applications en Théorie des Modules. V, 103 pages. 1972. DM 16,–

Vol. 255: Conference in Mathematical Logic – London '70. Edited by W. Hodges. VIII, 351 pages. 1972. DM 26,–

Vol. 256: C. A. Berenstein and M. A. Dostal, Analytically Uniform Spaces and their Applications to Convolution Equations. VII, 130 pages. 1972. DM 16,–

Vol. 257: R. B. Holmes, A Course on Optimization and Best Approximation. VIII, 233 pages. 1972. DM 20,–

Vol. 258: Séminaire de Probabilités VI. Edited by P. A. Meyer. VI, 253 pages. 1972. DM 22,–

Vol. 259: N. Moulis, Structures de Fredholm sur les Variétés Hilbertiennes. V, 123 pages. 1972. DM 16,–

Vol. 260: R. Godement and H. Jacquet, Zeta Functions of Simple Algebras. IX, 188 pages. 1972. DM 18,–

Vol. 261: A. Guichardet, Symmetric Hilbert Spaces and Related Topics. V, 197 pages. 1972. DM 18,–

Vol. 262: H. G. Zimmer, Computational Problems, Methods, and Results in Algebraic Number Theory. V, 103 pages. 1972. DM 16,–

Vol. 263: T. Parthasarathy, Selection Theorems and their Applications. VII, 101 pages. 1972. DM 16,–

Vol. 264: W. Messing, The Crystals Associated to Barsotti-Tate Groups: With Applications to Abelian Schemes. III, 190 pages. 1972. DM 18,–

Vol. 265: N. Saavedra Rivano, Catégories Tannakiennes. II, 418 pages. 1972. DM 26,–

Vol. 266: Conference on Harmonic Analysis. Edited by D. Gulick and R. L. Lipsman. VI, 323 pages. 1972. DM 24,–

Vol. 267: Numerische Lösung nichtlinearer partieller Differential- und Integro-Differentialgleichungen. Herausgegeben von R. Ansorge und W. Törnig. VI, 339 Seiten. 1972. DM 26,–

Vol. 268: C. G. Simader, On Dirichlet's Boundary Value Problem. IV, 238 pages. 1972. DM 20,–

Vol. 269: Théorie des Topos et Cohomologie Etale des Schémas. (SGA 4). Dirigé par M. Artin, A. Grothendieck et J. L. Verdier. XIX, 525 pages. 1972. DM 50,–

Vol. 270: Théorie des Topos et Cohomologie Etale des Schémas. Tome 2. (SGA 4). Dirigé par M. Artin, A. Grothendieck et J. L. Verdier. V, 418 pages. 1972. DM 50,–

Vol. 271: J. P. May, The Geometry of Iterated Loop Spaces. IX, 175 pages. 1972. DM 18,–

Vol. 272: K. R. Parthasarathy and K. Schmidt, Positive Definite Kernels, Continuous Tensor Products, and Central Limit Theorems of Probability Theory. VI, 107 pages. 1972. DM 16,–

Vol. 273: U. Seip, Kompakt erzeugte Vektorräume und Analysis. IX, 119 Seiten. 1972. DM 16,–

Vol. 274: Toposes, Algebraic Geometry and Logic. Edited by. F. W. Lawvere. VI, 189 pages. 1972. DM 18,–

Vol. 275: Séminaire Pierre Lelong (Analyse) Année 1970–1971. VI, 181 pages. 1972. DM 18,–

Vol. 276: A. Borel, Représentations de Groupes Localement Compacts. V, 98 pages. 1972. DM 16,–

Vol. 277: Séminaire Banach. Edité par C. Houzel. VII, 229 pages. 1972. DM 20,–